AGRICULTURAL MICROBIOLOGY

AGRICULTURAL MICROBIOLOGY

LAB TO LAND

By

Dr. Harsha Sharma
Assistant Professor
Motherhood University, Roorkee
Haridwar, (Uttarakhand)

DPH

DISCOVERY PUBLISHING HOUSE
INDIA

Published by:

DISCOVERY PUBLISHING HOUSE
4383/4B, Ansari Road, Darya Ganj
New Delhi-110 002 (India)
Phone : +91-11-23279245; 23253475; 43596065
E-mail : discoverybooksindia@gmail.com
discoverypublishinghouse@gmail.com
namitwasan9@gmail.com
web : www.discoverypublishinggroup.com

First Edition: **2023**

ISBN: 978-81-959169-0-0

Agricultural Microbiology

DEDICATION

I am expressing heart-felt gratitude to my beloved parents who always support me, motivate me at all stage and be with me with my first step.

Thanks to all those people who change the lives of others by living as an example and by empowering others for greatness.

Dr. NARENDRA SHARMA
Vice-Chancellor

MOTHERHOOD UNIVERSITY, Roorkee
ENLIGHTENING WORLD

Recognized by the UGC with the right to award degrees u/s 22(1) of the UGC Act 1956 and established under Uttarakhand Govt.Act No.05 of 2015

FOREWORD

I am happy to write a foreword to the first edition of "AGRICULTURAL MICROBIOLOGY" by Dr. Harsha Sharma, Assistant Professor, Faculty of Science, Motherhood University, Roorkee.

Agricultural microbiology is a branch of microbiology dealing with plant-associated microbes and plant and animal diseases. It also deals with the microbiology of soil fertility, such as microbial degradation of organic matter and soil nutrient transformations. Microbiology is considered to be one of the most applied branches of biology. Which are comparatively difficult to study in higher plants and animals. Some of the strange characteristics, survival of bacterial spores, the concept of enzyme energetics and metabolism make it more fascinating. Microorganisms affect positively and negatively on living beings. Microbiology includes various fields like vitamins, amino acids, agriculture microbiology, industrial microbiology and the production of several important chemicals besides antibiotics it is at large scale as well as small scale in the industry.

This book provides a chapter on the study of agricultural microbiology, microorganisms help in organic farming, environmental impacts of organic farming, pest and pathogen regulation in organic versus, research and development in agriculture, general principles in biodynamic farming, and economic management in organic farming.

The book provides the latest information in the field and will be informative and useful for all readers. I must congratulate to Author for their sincere efforts.

October 10, 2022

Prof. (Dr.) Narendra Sharma
Vice-Chancellor
Motherhood University, Roorkee
Uttarakhand, India

Roorkee-Dehradun Road, Vill. Karoundi, Post Bhagwanpur, Tehsil Roorkee, Distt. Haridwar (U.K.) India PIN - 247661

7060800160 vc@motherhooduniversity.edu.in www.motherhooduniversity.edu.in

Preface

Microbiology is a well-established biological science that is linked to a variety of professions in the modern day. Agricultural microbiology is a burgeoning scientific topic that emerged from the convergence of general microbiology and microbial ecology to agricultural biotechnologies. The ultimate goal of agricultural microbiology is to conduct broad research of beneficial bacteria and fungi interacting with agriculturally significant plants in order to meet world food demand in an environmentally sustainable manner. Our goal is to light a fire under the current state and application of microbiology in modern agriculture.

The present chapter gives an overall view about the role of microbiology in sustainable agriculture and also discusses effective microorganisms, rhizosphere, mycorrhizal fungi, phosphate solubilizing bacteria, impact of microbes on soil properties, synthetic biology, microbes as elicitors, microbes in stress agriculture, and climate smart agriculture.

Describe and evaluate major components of agricultural microbiology, such as soil fertility management, plant and animal production, social and environmental issues, as well as training and research, maintain a global perspective by drawing on a multinational team of authors and referring to the widest available data in each section, combine in one volume the insights of international experts who have direct experience with the organic movement, from on-farm work and teaching to marketing and rural appraisal; and provide a unique and timely science-based resource for researchers, teachers, students, primary producers and others around the world.

The book brings together a range of experts with considerable experience in farming. The book is not intended to be a how-to manual for organic farmers or a set of production and marketing recommendations; nor are the reviews intended to be uncritical depictions of organic principles and practices. Instead, the evaluations offer an objective and comprehensive examination of the problems discussed. The goal is to analyses farming rather than just to advocate for it. Some of the reviews offered may be constrained by a lack of data. This book is intended to be a one-of-a-kind and indispensable resource that provides a wide range of valuable information, data, and perspectives on the fundamentals of agricultural microbiology at a time when the global community is becoming increasingly aware of the problems with our current agricultural practices and the importance of developing sustainable agricultural and systems for the long-term health of humanity and the biosphere as a whole.

–Author

Contents

1

An Introduction to Agriculture Microbiology

The non-stop use of chemical fertilizer and pesticides for enhancement of crop yield and on the spot action of pest control causes harmful and dangerous outcomes on the surroundings because of the buildup of minerals and biomagnifications at the better hierarchical stage. Therefore, the modern need for opportunity and eco-friendly generation as incorporated Pest Management, plant boom-promoting microorganisms and organic be counted for reinforcing natural farming. Numerous agricultural soils worldwide are poor in plant vitamins. Hence, huge fertilizer requirement is a chief task for sustainable meals production. Previously, those plant nutrients were furnished completely within the shape of synthetic chemical fertilizers as chemical fertilizers are quite pricey and increase crop production fees. In addition, chemical fertilizer leads to soil degradation and fitness threat to both humans and farm animals. It is consequently imperative to improve soil fertility, whilst on the identical offering related poor environmental results of chemical fertilizer.

The organic count number consists of any plant or animal cloth that returns to the soil and is going via the decomposition method. In addition to imparting vitamins as a habitat to organisms living in the soil, the natural count number also binds soil particles into aggregates and improves the water-conserving capability of the soil. All over the world, humans tend to live in places with high fertile soils, enough rainfall, and appropriate temperatures. With population strain, soil nutrients and fertility lower while farmers are not capable of restriction losses by way of including nutrients to the soil through manure, crop residues, and biofertilizers in addition to mineral fertilizers. But the mineral fertilizers pose a health danger and have an effect on the microbial population in soil by degrading the physical structure of the soil, main to the lack of oxygen in the plant root zone besides being pretty expensive and making the fee of production excessive.

Different sources of biofertilizers such as nitrogen fixers, plant stimulators, phosphorus solubilizing microorganisms, and plant increase-selling rhizobacteria (PGPR) were pronounced; using bio-fertilizers became a totally vital eco-friendly practice and helped in getting amazing yield. Microorganisms in biofertilizers usually have particular capabilities; like Pseudomonas fluorescens fixes nitrogen and phosphate solubilizing bacteria solubilize phosphorus from the soil and make their smooth availability to the flowers. Nitrogen and phosphorus are the foremost nutrients for plant growth, although plants have a confined potential to take out them from the surroundings, consequently, they need microorganisms concerned with nutrient recycling, to assist flora uptake and absorption of these nutrients at adequate concentrations, while microbes get waste merchandise as food. During this symbiotic courting, plant rooting structures turn out to be stronger and bigger. As the plant root is bigger, greater living space and food are to be had for microorganisms, in a manner that microorganisms act as biofertilizers. Improvement of soil fertility through the use of biofertilizers like PGPR, mycorrhizal fungi have proven an essential contribution to nutrient availability in the soil, and also hold collectively soil debris into the solid aggregates, which bring down soil erosion and ameliorate soil structure. Different research has been conducted to evaluate the responses of diverse vegetable vegetation like chili, brinjal, and pea to biofertilizer software but findings have been still inconsistent.

As with all plant vegetation, nitrogen in the course of growth and grain development is important to facilitate protein accumulation inside the wheat kernels. Preparing biofertilizers from natural matters is practiced hardly ever. Bacteria (Pseudomonas fluorescens) colonizing roots that elicit shoot and root boom, referred usually as PGPR, are endorsed (marketed) for improving plant increase and disease manipulation both in agriculture and horticulture. PGPR can sell root and shoot growth either by way of producing plant hormones or secondary metabolites, controlling diseases, induction of systemic resistance, or converting physicochemical interactions with flora. Bacterial inoculants implemented as bio-fertilizers are also being explored to alleviate stress from biotic elements, for example, drought and salinity.

Some soils can suppress the severity of disease even within the presence of a pathogen, host plant, and favorable climatic situations for the disorder. Culture-primarily based and lifestyle-unbiased analyses have indicated the involvement of a diverse range of microorganisms involved in decreasing pathogen inoculums and contamination, plant boom promotion, and induction of systemic resistance. Due to the problems related to introducing inoculants, in situ enhancement of useful microorganisms involved in herbal disease suppression can be a greater effective and reliable manipulation measure. It also can provide environmental blessings with the aid of reducing agrochemical dependency for sickness manipulation. Organic farming promotes the use of crop rotations and cover vegetation, and encourages balanced host/predator relationships. Organic residues and nutrients produced on the farm are recycled lower back into the soil. Cover vegetation and composted manure are used to keep soil natural rely on and fertility. Preventative insect and disorder manipulation techniques are practiced, including crop rotation, improved genetics, and resistant varieties. Integrated pest and weed management, and soil conservation structures are treasured equipment on a natural farm. Organically authorized pesticides consist of "herbal" or other pest management merchandise included within the authorized substances listing of the organic requirements. The accepted substances listing identifies substances accredited for use as insecticides in Organic Farming. All grains forages, and protein supplements fed to cattle ought to be organically grown. With the boom in population, our compulsion would be now not handiest to stabilize agricultural manufacturing but to increase it further in a sustainable manner.

The scientists have realized that the – Green Revolution with high input use has reached a plateau and is now sustained with diminishing return of falling dividends. Thus, a natural balance needs to be maintained at all cost for existence of life and property. It may also cost heavily on our foreign exchange in future. In organic production, farmers choose not to use some of the convenient chemical tools available to other farmers. Design and management of the production system are critical to the success of the farm. Select enterprises that complement each other and choose crop rotation and tillage practices to avoid or reduce crop problems. Yields of each organic crop vary, depending on the success of the manager. During the transition from conventional to organic, production yields are lower than conventional levels, but after a three to five-year transition period the organic yields typically increase.

Cereal and forage crops can be grown organically relatively easily to due to relatively low pest pressures and nutrient requirements. Soybeans also perform well but weeds can be a challenge. Corn is being grown more frequently on organic farms but careful management of weed control and fertility is needed. Meeting nitrogen requirements is particularly challenging. Corn can be successfully grown after forage legumes or if manure has been applied. Markets for organic feed grains have been strong in recent years. Fruit and vegetable crops present greater challenges depending on the crop. Some managers have been very successful, while other farms with the same crop have had significant problems. In organic production, farmers choose not to use some of the convenient chemical tools available to other farmers. Design and management of the production system are critical to the success of the farm. Select enterprises that complement each other and choose crop rotation and tillage practices to avoid or reduce crop problems. Yields of each organic crop vary, depending on the success of the manager. During the transition from conventional to organic, production yields are lower than conventional levels, but after a three to five-year transition period the organic yields typically increase. Cereal and forage crops can be grown organically relatively easily to due to relatively low pest pressures and nutrient requirements. Soybeans also perform well but weeds can be a challenge. Corn is being grown more frequently on organic farms but careful management of weed control and fertility is needed. Meeting nitrogen requirements is particularly challenging. Corn can be successfully grown after forage legumes or if manure has been applied.

Markets for organic feed grains have been strong in recent years. The adoption of genetically engineered (GMO) corn and canola varieties on conventional farms has created the issue of buffer zones or isolation distance for organic corn and canola crops. Farmers producing corn and canola organically are required to manage the risks of GMO contamination in order to produce a "GMO-free" product. The main strategy to manage this risk is through appropriate buffer distances between organic and genetically engineered crops. Cross-pollinated crops such as corn and canola require much greater isolation distance than self-pollinated crops such as soybeans or cereals. An environmentally sustainable system of agriculture like organic farming will be able to maintain a stable resource balance, avoid over exploitation of renewable resource, conserving inherent soil nutritional quality and soil health, and biodiversity. It will lead us to sustainable agriculture and create a sustainable lifestyle for generations to come. The movement started with developed world is gradually picking up in developing countries. But demand is still concentrated in developed and most affluent countries. Local demand for organic food is growing. India is poised for faster growth with growing domestic market. Success of organic movement in India depends upon the growth of its own domestic markets. India has traditionally been a country of Organic Farming, but the growth of modern scientific, input intensive agriculture has pushed it to wall, with the increasing awareness about the safety and quality of foods, long term sustainability of the system and accumulating evidences of being equally productive, the organic farming has emerged as an alternative system of farming which not only addresses the quality and sustainability concerns, but also ensures a debt free, profitable livelihood option.

Agriculture's Development

There are currently around 7 billion people in the world and this is expected to increase to approximately 8 billion sometime around the year

2020. When one considers both the expected worldwide population increase and the increasing environmental damage that is a consequence of ever greater levels of industrialization, it is clear that in the next ten to twenty years it will be a significant challenge to feed all of the world people, a problem that will only increase with time. There is absolutely no time to lose; to feed this growing population, the world needs to begin to greatly increase agricultural productivity, and to do so in a sustainable and environmentally friendly manner. To feed the growing world, it is necessary to re-examine many of the existing approaches to agriculture that includes the use of chemical fertilizers, herbicides, fungicides, and insecticides. Instead, sustainable agriculture will likely make much greater use of both transgenic plants and plant growth-promoting bacteria, or PGPB. Green Revolution (GR) technologies, supported by policies, and fueled by agrochemicals, machinery and irrigation, are known to have enhanced agricultural production and productivity.

While these technologies greatly helped to address the food security of India, farmers using these technologies have to depend upon the purchased inputs. The small farmers, who by cash flow definition are short of cash, are therefore found to lag behind large farmers in the adoption of technologies. Modern agricultural farming practices, along with irrational use of chemical inputs over the past four decades have resulted in not only loss of natural habitat balance and soil health but have also caused many hazards like soil erosion, decreased groundwater level, soil Stalinization, pollution due to fertilizers and pesticides, genetic erosion, ill effects on environment, reduced food quality and increased the cost of cultivation, rendering the farmer poorer year by year. Farmers do not find agriculture a viable proposition anymore and in fact, a large number of farmers have committed suicides. Some of the factors that contributed to the present crisis in farming could be the shooting-up of the price of factory made external inputs and the government's slow withdrawal of investment as well as market intervention and more significantly, shifting of subsistence farming (mainly with homegrown inputs) to commercial farming (largely with purchased inputs). In other words, local indigenous farm techniques have been wiped out and replaced by the modern techniques, resulting in an unviable and unsustainable farm enterprise. It is in this context that alternative farm techniques and strategies for growing crops ought to be found in the larger interest. Soil scientists define soil microorganisms as any organism in soil which requires a microscope to observe. These organisms range from bacteria to nematodes and include diverse species of alga and protozoa. Soil microorganisms are responsible for the breakdown of organic matter, including hydrocarbons, conversion of organic components from one from to another and the production of humus.

Although this group is large and diverse, soil microorganisms from a robust community capable of surviving and functioning under extremes of temperature, water availability, soil pH, energy resources, nutrient availability and salt concentration.

The principle of organic cultivation is attracting farmers world over due to its various advantages over modern agricultural practices. Essentially, it is a farming system which supports and strengthens biological processes without recourse to inorganic remedies such as chemicals or genetically modified organisms. Organic Farming is productive and sustainable. Many state-supported agencies, non-governmental organizations (NGOs) and individuals have started experimenting with organic methods of food production in the recent past. The most popularly accepted definition of organic farming is: Organic Farming is a holistic production management system which promotes and enhances agro ecosystem health, including biodiversity, biological cycles, and soil biological activity, it emphasizes the use of management practices in preference to the use of off-farm inputs, taking into account that regional conditions require locally adapted systems. This is accomplished by using wherever possible, agronomic, biological and mechanical methods, as opposed to using synthetic materials, to fulfill any specific function within the system.

The term **'conventional farming'** refers to a production system which employs a full range of pre-and post-plant tillage practices (e.g., plough, disc plant, and cultivator), synthetic fertilizers and pesticides. It is characterized by a high degree of crop specialization. In contrast, organic farming is characterized by a diversity of crops. In world, the increasing population on a near stabilized agricultural land places a heavy burden on the soil resource particularly its nutrient supplying power. Intensive agriculture involving the use of chemical fertilizers in large amount has no doubt, resulted in manifold increase in the productivity of farm commodities but, adverse effects of these chemicals are clearly visible on soil structure, microflora, and quality of water, food and fodder.

A Modern Organic Movement

Scientists became increasingly interested and aware of Organic Farming in the 1980s, even those who were not supportive of alternative agricultural systems. They found the academic climate and funding sources were more amenable to its study than in previous decades, which resulted in a rash of research, much of which, unfortunately, was comparisons of organic and non-Organic Farming, rather than research designed to assist organic producers or underpin organic principles and practices. By the end of the decade, the level of interest in Organic Farming and the volume of information compiled about organic methods had become sufficient to enable the highly successful publication of the landmark book Organic Farming.

Trends that began in the 1970s, and accelerated through the 1980s, continued to flourish during the 1990s and into the new millennium. Demand and production continued to grow exponentially around the world, often at 20-30% per year. Formal political and legislative recognition was achieved. Normally this was started by bringing Organic Farming under legislative control. Following this were intergovernmental agreements to facilitate international organic trade, mostly by creating systems by which certification standards in the exporting country were shown to be equivalent to those of the importing country, a system that parallels and duplicates IFOAM's Organic Guarantee System. Significant political traction was also being made in international/intergovernmental agencies such as the European Union (EU) and the United Nations Food and Agriculture Organization. Public concerns about food and its production systems continued with further food scares such as bovine spongiform encephalopathy (BSE) in the UK and the emergence into public awareness of genetic engineering that in parts of asia, science increasingly became a tool to demonstrate the benefits of Organic Farming and the problems with industrial agriculture. This helped organic organizations make the case for much closer cooperation between themselves and other environmentally aligned organizations, for example nature conservation groups. It also showed that useful research could be carried out on organic farms. Since the 1980s, numerous organic research centers and associations have been established internationally; taken active roles in conducting new research in the agronomic, environmental and social sciences; have documented and published findings to fill the strong demand for information; and provided extension and training to farmers and advisers. Several NGOs and companies began to perform an auxiliary function to the certifying agencies by carrying out independent reviews of products intended for use in certified organic production, handling and processing.

By the late 1990s increasing concerns were being raised about Organic Farming following in the footsteps of industrial agriculture and losing its vision. Examples of this are the huge growth in sales though supermarkets and increasing amounts of organic produce being transported large distances to satisfy demand in affluent countries. These concerns have resulted in a refocusing on the neglected issue of social equity (e.g., ensuring that farmers are paid a fair price for their produce). One outcome of this is the linkages formed between the Fair Trade and the organic movement. There is active debate on introducing Fair Trade requirements for European organic producers that have, to date, only been used by farmers in the third world. A practical example of reforming the links that existed between organic producers and consumers in the 1960s and 1970s are the rapid increase in farmers markets in the USA and UK where traditional produce markets have been resurrected by requiring stall holders to be both local and only sell goods they have produced.

2

The Agricultural impact of Sustainable Agricultural Practices

Agricultural production currently depends on the large-scale use of chemical fertilizers. These fertilizers have become essential components of modern agriculture because they provide essential plant nutrients such as nitrogen, phosphorus and potassium. However, the overuse of fertilizers can cause unanticipated environmental impacts. To achieve maximum benefits in terms of fertilizer savings and better growth, the PGPB-based inoculation technology should be utilized along with appropriate levels of fertilization. Moreover, the use of efficient inoculants can be considered an important strategy for sustainable management and for reducing environmental problems by decreasing the use of chemical fertilizers. In Organic Farming is developing rapidly and today at least 141 countries produce organic food commercially. As per the estimates in the year 1999, organic food is produced in about 32.2 million hectares globally, managed by more than 1.2 million producers, including smallholders. In addition to agricultural land, there is 0.4 M ha of certified organic aquaculture. Among the countries involved in organic farming, about 65 per cent are developing countries.

The regions with the largest areas of organically managed agricultural land are: Oceania, Europe and Latin America. Australia, Argentina and Brazil are the countries with the largest organically managed land areas. About one-third of the worlds organically-managed land-almost 11 Mha - is located in the developing countries. Most of this land is in Latin American countries, while Asia and Africa take the second and third places, respectively. On a global level, in the year 1998, organic land area increased by almost 1.5 M ha compared to the data for the year 1999. About 28 per cent (or 1.4 Mha) more land under organic management was reported for

Latin. In Europe, organically 0.9 Managed land increased by 0.33 Mha (+ 4%) and by Mha (+27%) in Africa. Austria has the highest percentage (8.40%) of area under organic farming, followed by Switzerland, UK and Germany. In India, only 0.03 per cent of the area is under organic farming, though there is huge scope for bringing more land under organic farming. Agriculture based on chemical use and intensive irrigation has become apparent and there has been a resurgence of interest in Organic Farming. Renewed interest in Organic Farming is mainly due to two concerns, falling agricultural yield in certain areas as a result of inter alia excessive use of chemical inputs, decreased soil fertility and environmental awareness. Exports also played a role but perhaps lesser than in other countries.

The 10th Five-Year Plan encouraged the promotion of organic farming using organic wastes, and integrated pest management (IPM) and integrated nutrient management (INM) practices. Even the 9th Five-Year Plan had emphasized the promotion of organic produce in plantation crops, spices and condiments using organic and bio-inputs for the protection of environment and promotion of sustainable agriculture. Presently, many states and private agencies are involved in the promotion of organic farming in India; these also include several ministries and government departments at both central and state levels. The states of Uttarakhand and Sikkim have declared their states as organic states. In Maharashtra, since 1998, about 5 lakh ha area has been under organic farming (of the 1.8 crore ha of cultivable land in the state).

The Vidarbha Cotton Growers Association, set up in 1994 with 135 members, has tied up with international agencies for exports. In Gujarat, organic production of chickoo, banana and coconut were found to be more profitable, though field crops and mango had both lower input costs as well as yields. In Karnataka, 1513 ha land was under certified organic farming, and 4750 ha was under non-certified organic farming by the year 2000. The major reasons for shift towards organic farming include sustained soil fertility, reduced cost of cultivation, higher quality of produce, sustained yields, easy availability of farm inputs and reduced attacks of pest and diseases. The Government of Karnataka released a state organic farming policy in 1999. Most of the area in the north-eastern states is being used for organic farming. In Nagaland, 3000 ha area is under organic farming. States like Rajasthan, Tamil Nadu, Kerala, Madhya Pradesh, Himachal Pradesh and Gujarat are promoting organic farming vigorously. Organic farming is an intensive process, limited mostly to resource-rich farmers, and the export market and depends heavily on external support systems for price, market intelligence and certification of produce, among others. Hence, he has concluded that the scope of coverage and social relevance of the organic farming is also limited. Instead, he has proposed ecological farming whose main objectives are maintenance of high productivity, reduction in production cost and enhancement in self-reliance. It caters to both resource-poor and the resource-rich farmers; the process is simple, addresses local market and the scope of coverage and social relevance is also high. There are four main steps in ecological farming: the first being the adoption of non- chemical pest management methods; the second is to focus on selling pesticide- free produce in the local market; the third is establishment of community managed seed banks; and finally, the fourth step is to adopt non-chemical methods of nutrient management. It has been argued that the ecological method is indigenous but is gradually disappearing due to constraints in labour availability. There is a great scope for its revival by utilizing the incentives of labour under the National Rural Employment Guarantee Act (NREGA).

Association of India popularized the concept of Permaculture (permanent agriculture) in AP. Permaculture is the conscious design and maintenance of agriculturally productive ecosystems which have the diversity, stability, and resilience of natural ecosystems. The Deccan Development Society (DDS) an internationally well-known NGO working with dalit women groups has developed a farm on the principles of Permaculture in Zaheerabad region of deccan area. The DDS encourages sustainable agricultural practices in a big way and has been a pioneer in the country. More than 5000 women farmers in an area of more than 20,000 acres have adopted sustainable agricultural practices, which are environment friendly and are based on the traditional knowledge. Similarly, the Centre for Sustainable Agriculture (CSA) 2 based at Hyderabad, through several NGOs in the state, has promoted non-pesticidal management of pests in the state, wherein the use of pesticides and chemical fertilizers is discouraged, while the use of local resources is encouraged.

The Community Managed Sustainable Agriculture program is being implemented by the Society for Elimination of Rural Poverty (SERP), the Government of Andhra Pradesh and the Sustainable Agriculture Network of NGOs, with technical support from the Centre for Sustainable Agriculture. In 1999, there were 50 villages which had become pesticide free and 7 villages which have become completely organic. The Timbaktu Collective is another organization which has been promoting organic farming practices since a long time in the Ananthapur district. The Government of Andhra Pradesh has initiated programs related to organic farming through the Departments of Agriculture and Horticulture. The Agriculture Department of AP had taken up the promotion of organic farming in the state during the year 1999-2000 by implementing several schemes with an outlay of Rs 18.29 crore. Similarly, the Horticulture Department of AP is implementing the organic farming scheme under the State Horticulture Mission (SHM) from the financial year 1999-2000. The Andhra Pradesh states policy on organic farming is yet to be finalized and the draft developed in this regard is being discussed at various levels.

Farmers largely convert to organic farming because of the uneasiness experienced with the existing agriculture system, which is predominantly based on chemicals. Some farmers perceive chemical agriculture to be health hazard for themselves. However, personal health is not the only reason to convert to organic farming. Farmers in Punjab, Haryana and Eastern Uttar Pradesh are able to maintain their yield levels only through a drastic increase in chemical inputs. Yields in irrigated farms may go down during the conversion period from conventional to organic farming because the crop yields are boosted by artificial fertilizers and it takes time for the soil fertility to get boosted. However, after conversion, yields will be equal, if not higher than the yield during the conventional farming. In the rainfed farming, the situation is different; yields are significantly lower and thus, the difference in yields between the conventional and conversion period is narrow. Though comparative yield studies are only a few at both global and national levels, certain studies have provided a broad indication about the productivity of organic farms vis-à-vis conventional farms.

Conversion from the traditional low-external input system of cultivation rarely results in lower yields. However, when switching from external-input-intensive forms of agriculture, the yields may decline significantly, at least during the initial years of conversion, until the natural soil tilth and fertility are sufficiently restored. But, after that, they may stabilize at comparably, lower or even higher levels, depending on the efficacy of organic management and the quality of organic fertilizers applied. The wide range of organic fertilizers that are based on local resources and farmer's knowledge will take care of the manorial needs of farmers. Organic farming can compete economically with conventional farming when specific attention is given to optimum approaches while conversion. Information needs of organic farmers should be surveyed and information delivery systems should be tailored to meet those needs. In the case of crops like rice, organic cultivation appears to be less economical as compared to other crops. However, there is more scope for minimizing the economic cost and environmental loss, under organic farming system in the long-run. Besides these, environmental balance is maintained such that crops, trees, animals and man can live more harmoniously. Reducing the use of pesticide can provide the growers with direct economic benefits by decreasing the cost of inputs, thereby increasing net returns. Crop diversity in organic farms can have other economic benefits as diversity provides some protection from adverse price changes in a single commodity. Most organic farming practitioners have reported that it was not the premium price of the organic produce but the reduced expenditure on inputs and similar yields to their neighboring conventional farmers that motivated them towards organic farming.

Recently, based on their experiments going on for 25 years in Switzerland, have reported sustainable yields (though marginally reduced in some years) without agrochemicals in the temperate climatic conditions. On the other extreme, most agricultural scientists believe that without chemical fertilizers, large quantities of farm yard manure (FYM) and other biomass material that will be needed to compensate for the fertilizers are not unavailable. They also believe that there are several crops that cannot produce high yields without the use of agrochemicals, fertilizers in particular and therefore, practicing organic farming means food insecurity for the country. Organic farmers need to borrow less money than conventional farmers for two reasons; firstly, organic farmers need to buy fewer inputs such as fertilizer and pesticides; and secondly, costs and income are more evenly distributed throughout the year on diversified organic farms. Organic farmers, however complain that they face discrimination by lenders, a possible economic disadvantage of organic farming.

However, concluded that this problem is more perceived than real. Income and profitability of organic farms is equal or higher when compared to conventional and traditional farms. In the long run, organic farming offers more advantages compared to conventional farming, because it not only promises higher yields but also ensures higher yield security and reduces dependence on external inputs, thus making poor households less crisis-prone. These are weighty arguments, especially in the marginal locations have compared the economic performance of 14 organic crop/ livestock farms in the Midwest with that of 14 conventional farms. The farms under study were paired based on the physical characteristics and types of farm enterprises. The market value of crops produced per unit area was 11 per cent lower on the organic farms. But since the cost of production was also less, the net income per unit area was comparable for both the systems. Compared data from 15 organic farms in the western Corn Belt with the USDA data on representative conventional farms in the same area. In most cases, the net returns were higher on the organic farms. Both the studies have shown that production costs were lower on the organic farms.

Two studies comparing cash grain farms were conducted in the state of Washington. Made a comparison between three organic and three conventional farms and found that net income per unit area was 38 per cent higher on the conventional farms. However, the author of a follow-up study of six organic farms has found that net returns on organic farms were 22 per cent higher than on the representative conventional farms has compared 10 organic and 10 conventional farms in New York and Pennsylvania for returns from wheat production only. When cash operating costs alone were included, the returns were higher on the organic farms. However, when the costs of land and unpaid family labor were included, the conventional farms had a higher average net return. However, the above studies had several limitations. The most obvious was the small sample size, which made it difficult to conduct any statistical test of differences. The averages did not reflect the high variability that occurred in both yields and net returns on both types of farms. Pairing farms for the studies also caused problems, especially in the works by. Finally, none of the studies has included the livestock enterprise which may be essential for optimum economic performance of organic farms.

The sample may not have been a representative sample of organic farmers, and many of the responses may have been based on perceptions rather than on well-kept records. The survey seems to indicate a high level of satisfaction with the economic performance of low input farming. Organic biodiversity farming and food production is the way out for the ecological and livelihood security of millions of small farmers in this country. Organic farming benefits the society substantially by reducing pollution and conserving energy, soil nutrients, fish, wildlife and insuring the supply of food for the future generations. However, virtually no credible data are available to policymakers on the magnitude of these benefits; they are unable to compare organic farming with other policy alternatives. In areas where organic farming is known to be economically feasible, policy barriers to conversion should be identified, evaluated and addressed. Organic farming is an attractive alternative for both farmers and policymakers. The new biochemical technology in agriculture, however, has many negative impacts on the environment. There has been a significant increase in the use of chemicals like fertilizers and pesticides since the 1960s. There is enough cause for worry on the environmental consequences of these chemicals. Particularly in the 1980s, it was realized that for the sustainable development, alternative farming practices are needed. Continuous usage of pesticides application has led to diseases like cancer and epilepsy with which the people are being made to suffer for years.

Alternatives to pesticides are to be found viable in the long-run and hence, a concerted effort needs to be put by all concerned for promoting sustainable agricultural development (SAD) in the broader framework of environment and health. The economic and environmental impact of our farm policies on pesticide reduction also deserves scrutiny and policies that encourage adoption of ecologically sound farming practices need to be implemented. Although many trained farmers realize the importance of ecological agriculture, it is not always possible for them to put the training into practice, especially on their major farming land which provides them with most of their livelihood security. However, farmers have adopted this technique to a greater extent on their homestead land, which is less controlled by market forces and is free from other external factors. This perhaps reflects their belief in the need for such an approach. These findings do indicate that the level of awareness among farmers is rising significantly, though there is still a long way to go before there is a total shift from inorganic to organic farming.

It has been found that in places like Chhattisgarh where Organic Farming is popular, the farmers who try to practice Organic Farming suffer, as the upstream farmers may be using chemicals which permeate into the fields of farmers practicing organic cultivation and the produce would be found contaminated during chemical analysis due to the residual effect across the fields. This is more so in the case of medicinal plants, where the sensitivity index is much higher owing to their use in the life-saving drugs or health products. It has been found that the organic cultivation movement can become a success only when the farming community is jointly sensitized and mobilized to give up inorganic practices.

The major factors that lead to growing interest in the alternative forms of agriculture in the world are: increasing consciousness about conservation of environment as well as health hazards associated with agrochemicals, and consumer's preference to safe and hazard free food. Organic Farming is one among the broad spectrum of production methods that are supportive of the environment. The demand for organic food is increasing steadily in both the developed and developing countries at an annual average growth rate of 20-25%. Considering the potential environmental benefits of organic production and its compatibility with integrated agricultural approaches to rural development, Organic Farming may be considered as a development vehicle for the developing countries such as India. A comparative study on economics of crop production under Organic Farming System (OFS) and Inorganic Farming System (IFS) showed that production cost was gradually declining in OFS. Further, it is not easy to assign economic values for soil health, reduced pollution and improved resilience and reduced greenhouse gas emissions. Changes in soil structure coupled with improved ground cover, decreased runoff by about 10 to 50 per cent and increased infiltration by about 10 to 25 per cent, all these factors combined to reduce soil erosion on organic fields by at least two-fifths, and sometimes, over four- fifths. It is difficult to place a monetary value on the water lost as runoff and the nutrients contained in the eroded soil. In part, they are just displaced to other locations on the farm, where they remain available for crop production.

Despite serious efforts of some NGOs, it appears that India is lagging far behind in the adoption of organic farming. For laying the spadework for the spread of Organic Farming in the country, certain issues require attention at the government policy making levels. These include (a) substantial financial support by the governments which is absolutely necessary to promote organic farming; (b) market development for the organic products which is a crucial factor to promote domestic sales; (c) government support to the producer and consumer associations to market the organic products; (d) simplification of the process of certification; and (e) reduction in certification cost. A vigorous campaign to highlight the benefits of organic farming against the conventional system is essential to increase awareness of both farmers and consumers. There is no mention of organic farming in the National Agricultural Policy. Organic farming offers an alternative method for production that can be suitably exploited to benefit some segment of farmers. However, certification of organic products becomes dubious if it is linked with high documentation, controlling, organizational and bureaucratic effort. In Chhattisgarh, through various initiatives, the government has been promoting the cultivation of medicinal, aromatic and dye plants, apart from agricultural and horticultural produce. Being an herbal state, agencies there is a lot of scope for promoting organic farming. The Chhattisgarh Vanoushadhi Board or the Medicinal Plants Board, the Departments of Horticulture and, Agriculture, and Chhattisgarh State Minor Forest Produce Federation are some of the state governments promoting organic cultivation of agricultural, horticultural, medicinal and aromatic crops. It has suffered downright neglect by the central government, a number of state governments have already made significant strides in organic farming. The governments of the mountainous states of Sikkim, Mizoram and Uttarakhand have undertaken significant initiatives to turn their states completely organic. State government initiatives in some forms have also been taken in Karnataka, Madhya Pradesh, Arunachal Pradesh, Meghalaya and Punjab. In the Uttarakhand organic initiative, a multi-pronged strategy-the organic model-has been promoted not only as an agricultural technology, but also as an integral part of several rural development projects. Moreover, while export is not outside the purview of this initiative, significant emphasis has been placed on domestic market development as well. Although it is too early to comment on this programme, it seems that if implemented successfully, the project could become a role model for state-driven organic development in India. India is endowed with various types of naturally viable organic form of nutrients across different regions of the country which will be helpful in organic cultivation of crops. This will help substantially in organic cultivation of crops. There is a wide diversity in climate and eco-system. India has a strong traditional farming system with innovative farmers, vast drylands and least use of chemicals. Infect, the rainfed tribal, north-east and hilly regions of the country where negligible chemicals are used in agriculture, have been practicing subsistence agriculture for a long period; such areas are organic by default.

Pests and Diseases in Organic and Conventional Agriculture

Pests and diseases that plague conventional farming operations, causing yield loss or the application of costly inputs, are often the same species that challenge organic growers producing the same crops. One significant difference is that organic growers avoid the use of broad-spectrum synthetic pesticides, which severely disrupt natural controls in the system and promote the occurrence of secondary pests. Well-known secondary pests in pesticide intensive systems include spider mites in temperate orchards treated for codling moth, rice brown plant hopper in pesticide- treated tropical paddy rice, Rhizoctonia black scurf in potato after nematicide applications that reduce fungi feeding collembola, and apple scab as a result of decimation of earthworm populations by the fungicide benomyl (slowing down decomposition of infected leaves). A substantial number of major pests in conventional systems, then, are regulated at low levels in organic systems by virtue of the conservation of their natural enemy complex.

Natural pest and pathogen controls are not only conserved (not disrupted) but are also promoted in organic farming conditions. Most soil borne plant pathogens causing root and foot rots in older plants are usually less prevalent in organic than in conventional farms. This kind of disease suppression has frequently been associated with higher microbial activity and diversity, with higher macrofaunal numbers and diversity, and/or with lower soil and crop N concentrations in organic than in conventional soils. Attacks by some airborne diseases (in particular many powdery mildew and rust diseases) and by sucking insect pests (aphids and whiteflies) can also be less severe in organic than in conventional crops due to lower nitrogen concentrations in foliar tissues or phloem on organic than on conventional farms. Some arthropods are favored under conditions of organic farming practices, however, particularly below-ground pests that are fostered by rich organic matter such as the garden symphylan, cutworms, wireworms and slugs or hardy pest insects that have few biological controls and are not effectively controlled with allowable organic inputs such as the strawberry weevil or Lygus bug.

Similarly, damping-off causing pathogens such as Pythium species can wreak havoc in organic crops, since these can multiply quickly in fresh organic materials incorporated into soil. Certain foliar diseases that can spread quickly and are controlled by frequent fungicide sprays in conventional farms, such as potato late blight and onion downy mildew, can be devastating on organic crops in humid climates. Stored products pests should also be a particularly challenging problem for Organic Farming, since synthetic insecticides are prohibited. However, it seems that stored products pests are a universal problem posing challenges for conventional and organic farmers alike. Many problems with synthetic chemical approaches in the past decades to long-term storage pest control, and propose new innovations in alternative control strategies. These control strategies include those used by organic growers. Vertebrate pests, such as deer and other ungulates, fruit eating and seed-eating birds, rodents, rabbits and squirrel's colonies or intermittently visit organic and conventional farms, potentially reducing yields and/or affecting food quality. Some of the practices more common on organic farms such as cover cropping, farm scaping with non-crop vegetation, and mixed cropping encourage beneficial fauna and repel some vertebrate pests, but may also improve the habitat for other vertebrates, such as gophers, voles and noxious birds. Organic practices overlap with those of conventional farming for control of these pests, including sanitation, exclusion, trapping and the use of a variety of repellents (visual, vegetation management, auditory) but exclude the use of many fumigants, anticoagulants and toxins.

Organic growers may therefore expect to encounter most of the same pests and pathogens as conventional growers in their fields. However, the dynamics of these organisms depend on the extent to which the organism's resource requirements are met and degree to which natural controls are functioning in the management system. Therefore, for many pests and pathogens, organic practices reduce the probability of establishment and spread in the crop field despite the presence of the same crop host. For other pests and pathogens, organic practices promote suitable conditions for population growth, resulting in particular challenges for growers with limited agrochemical options.

Pest and Disease Differences in Organic and Conventional Agriculture

Crop protection in Organic Farming is accomplished through three general approaches, by:

1. Preventing colonization by pests and pathogens;
2. Regulating the abundance of pests and pathogens at low levels through biological processes; and
3. Employing curatives that are allowed under Organic Farming guidelines.

The first two approaches tend to be prophylactic control measures and the latter involves the substitution of synthetic chemical inputs used in conventional crops once pest or disease levels begin to rise. Because prophylactic crop protection in organic systems is based on ecological processes, the crop is seen as a host, the crop field and surroundings as a biotic community with it's a biotic condition, and the pest or pathogen as an invader that colonizes the crop habitat, establishes and erupts. Barriers to pest outbreaks include various means of isolating the crop from pest/ pathogen source pools and regulating pest and pathogen populations once established through community resistance. Community resistance comprises those factors that cause the habitat to be unsuitable for proliferation of invaders, such as resource limitation, competition, and predation. Community resistance compensates for synthetic pesticides in conventional agriculture. Above - ground, community resistance involves the conservation and enhancement of beneficial fauna, either directly or via diversified vegetation.

In soil, community resistance can be enhanced by activation of the soil food web through amendment with slowly decomposing organic materials. Enhancement of biodiversity is the key element in these efforts. Curative measures, taken as pest or pathogen populations begin to rise, include application of organically approved biocides or behavior modifying compounds derived from natural sources or inundate releases of other organisms (competitors, predators, or parasites). Curative measures may substitute for synthetic pesticides in conventional agriculture, and can complement other tactics, such as biodiversity enhancement. Successful crop protection in organic systems relies on prophylactic measures that prevent pest and pathogen colonization, establishment or build up sometimes in combination with curative measures when needed. Prevention of colonization or establishment of pests and pathogens in Organic Farming Colonization of the crop field, orchard, vineyard, storage facility or other agricultural environment by pests and pathogens is prevented through sanitation, source isolation, and other protective measures. Practices to prevent colonization and establishment of pests and pathogens, namely sanitation, clean seeds or vegetative propagating materials, crop rotation, adjustment of planting time, removal of certain weeds, fencing or netting against vertebrates, sealing or repelling against storage pests, hold for both organic and conventional agriculture. However, they are even more important for organic farming, because curative measures are restricted here. The use of various crop protection practices to prevent colonization of the crop by pests and pathogens are at least as common in Organic Farming as in conventional agriculture. To illustrate some of the particular problems facing organic growers, we consider seed sanitation and crop rotation.

Pathogens and pests can also be avoided by adapting the crop planting time or rotating crops that harbor different suites of pests and pathogens (temporal isolation). The time between rotations of a particular crop is usually longer in organic than conventional field crop production (5-8 years-23 years). Successive planting of the same crop is not permitted according to organic standards in the EU, but rotation times are minimized in organic greenhouse production as only high value crops with similar cultural practices can be grown such as tomato, sweet pepper and cucumber in rotation. Organic growers practicing short crop rotations in greenhouses can face severe problems with root knot nematodes in soil-bound production of these high-value crops.

Resistance of host plants When the following conditions are met, the quality of the host plant is optimized for crop protection:

1. Adequate nutrient status for plant productivity and health is maintained, without excess nutrients or imbalances that support high levels of herbivores or pathogens; and
2. Toxic or repellent properties are sufficient to directly reduce pest or pathogen exploitation and survival.

The first approach is somewhat flexible, allowing for a grower to respond to pest and pathogen dynamics. However, the decision to use a resistant variety is set for the season. Its use will be determined by the probability of invasion, the severity of the pest or pathogen, any associated loss of yield quality or quantity, marketability, complementarily with other crop protection tactics, and the effectiveness of the resistant cultivar against the target and other possible exploiters. A caveat to these generalizations is that one organism's famine is another organisms feast, that is, very high nitrogen may attract an insect and deter a plant parasitic nematode (if ammonia is released from the nitrogen source). High mustard oil content deters insects that find it toxic, and attracts those that specialize on mustard family crops.

Organic System Studies

There is a need for rethinking the approaches, processes and institutional structures of agricultural research, because of the range and scale of consequences that agricultural research is expected to address today. There are high political demands on the relevance and proactive perspective of research relative to the changing goals, intentions and values of society and agriculture. These demands are not restricted to agricultural research. They are part of a more general change in the conception of science and its role in society, from that of science as an independent source of objective knowledge to that of science as special learning process within society. Agriculture is an area undergoing rapid development, both in terms of technological development and the development of alternative production systems, and agricultural research is influential in these developments. In this sense, agricultural science is a systemic science, a science that influences its own subject area. Furthermore, agricultural practice involves both social and ecological systems, and research into these socioecological systems faces the dual challenge of understanding complex agro ecosystem interactions and the practices of people in social systems. Agricultural systems research is, therefore, inherently framed in a social context, and necessarily involves questions concerning different interest and values in society as well as different structures of rationality and meaning.

There is, therefore, a need to explicitly address how values in the form of intentions and social interests feature in agricultural research. This is in terms of where and how values enter into the research process, and also in terms of what the systemic nature of agricultural research means in relation to the conventional scientific criteria of quality. This need applies generally to agricultural research. Moreover, the role of values is particularly evident with regard to organic farming, because special values and goals are obvious and decisive and because these values are clearly different from the values of mainstream agriculture. The special values of Organic Farming are summarized and what this obvious and manifest value basis means for organic research, how is to be performed, and how it is to be evaluated is briefly considered. Sound research approaches and appropriate methodologies for organic research are described and the relationship between organic and conventional research is discussed.

Standards, Principles, and Ethics

Organic farming has differentiated itself from conventional agriculture by way of alternative agricultural practices, world views and values. Most notably, the organic movement has explicitly formulated basic principles and standards for organic production and processing. The principles are based on a perception of humans and human society as an integrated part of nature and a holistic conception of health. Understanding the ecological processes that drive productivity and environmental impact through soil biology, vegetation dynamics, pest population dynamics, disease epidemiology and so on, is a key to improved organic systems.

This does not imply that they are unimportant in conventional farming. Organic farming does, however, not have access to all the technological means for overcoming natural and ecological insufficiencies and problems (notably artificial pesticides, fertilizers, preventive medicine) that conventional farming has, and it therefore relies on cooperation with natural ecological systems to a greater extent. Values of animal welfare, biodiversity and livelihood are also an integral feature of organic farming. This comprehensive, integrated view of nature and ethics is evident in one of the original key ideas of Organic Farming that the health of soil, plant, animal and man is one and indivisible.

In the existing IFOAM Basic Standards (IFOAM 2002), there are 15 principal aims plus general principles for each area in the standards, and the principles are thus quite complex to overview. There is, however, a process going on within IFOAM to rewrite the principles of Organic Farming. The process is expected to result in a set of basic ethical principles to be decided upon in the General Assembly 2005 in Adelaide. In the draft of May 2005, there are four principles on health, ecology, fairness and care. From the perspective of research, the formulation of basic ethical principles of Organic Farming have been advocated as a necessary tool for researchers to initiate far-seeing, proactive research that can assist the development of Organic Farming, and the sustainable development of agriculture in general, in a constructive and critical way. The present IFOAM process will hopefully provide a simpler and more consistent source of organic principles for researchers to use as a starting point when planning and performing research on Organic Farming.

Organic Research Quality Criteria

In the same way that proponents of conventional agriculture have criticized organic farming so too have conventional researchers criticized organic research. This kind of critique can lead the organic research community to reflect on the quality of past organic research, and whether organic research can be of high quality. The answers to these questions depend entirely criteria of quality that are used. Some of the critiques of organic research are based on the presumption that science should be free of values and ideologies, has termed ethical reductionism. The argument is that Organic Farming is inherently ideological, science should operate without reference to ideologies and, therefore, research that operates from the viewpoint of organic ideology and values is not scientific.

Organic Farming Methods Include

Organic farming, agricultural system that uses ecologically based pest controls and biological fertilizers derived largely from animal and plant wastes and nitrogen-fixing cover crops. Modern organic farming was developed as a response to the environmental harm caused by the use of chemical pesticides and synthetic fertilizers in conventional agriculture, and it has numerous ecological benefits. Compared with conventional agriculture, organic farming uses fewer pesticides, reduces soil erosion, decreases nitrate leaching into groundwater and surface water, and recycles animal wastes back into the farm. These benefits are counterbalanced by higher food costs for consumers and generally lower yields. Indeed, yields of organic crops have been found to be about 25 percent lower overall than conventionally grown crops, although this can vary considerably depending upon the type of crop. The challenge for future Organic Farming will be to maintain its environmental benefits, increase yields, and reduce prices while meeting the challenges of climate change and an increasing world population.

The concepts of Organic Farming were developed in the early 1900s by Sir Albert Howard, F.H. King, Rudolf Steiner, and others who believed that the use of animal manures (often made into compost), cover crops, crop rotation, and biologically based pest controls resulted in a better farming system. Howard, having worked in India as an agricultural researcher, gained much inspiration from the traditional and sustainable farming practices he encountered there and advocated for their adoption in the West. Such practices were further promoted by various advocates such as J.I. Rodale and his son Robert, in the 1940s and onward, who published Organic Gardening and Farming magazine and a number of texts on organic farming. The demand for organic food was stimulated in the 1960s by the publication of Silent Spring, by Rachel Carson, which documented the extent of environmental damage caused by insecticides. Organic food sales increased steadily from the late 20th century. Greater environmental awareness, coupled with concerns over the health impacts of pesticide residues and consumption of genetically modified (GMO) crops, fostered the growth of the organic sector.

Organic Farming Methods Fertilizers

Since synthetic fertilizers are not used, building and maintaining a rich, living soil through the addition of organic matter is a priority for organic farmers. Organic matter can be applied through the application of manure, compost, and animal by-products, such as feather meal or blood meal. Due to the potential for harboring human pathogens, the USDA National Organic Standards mandate that raw manure must be applied no later than 90 or 120 days before harvest, depending on whether the harvested part of the crop is in contact with the ground. Composted manure that has been turned 5 times in 15 days and reached temperatures between 55-77.2° C (131-171° F) has no restrictions on application times. Compost adds organic matter, providing a wide range of nutrients for plants, and adds beneficial microbes to the soil. Given that these nutrients are mostly in an unmineralized form that cannot be taken up by plants, soil microbes are needed to break down organic matter and transform nutrients into a bioavailable mineralized state. In comparison, synthetic fertilizers are already in mineralized form and can be taken up by plants directly. Soil is maintained by planting and then tilling in cover crops, which help protect the soil from erosion off-season and provide additional organic matter. The tilling in of nitrogen-fixing cover crops, such as clover or alfalfa, also adds nitrogen to the soil. Cover crops are commonly planted before or after the cash crop season or in conjunction with crop rotation and can also be planted between the rows of some crops, such as tree fruits. Researchers and growers are working to develop organic farming no-till and reduced-tillage practices in order to further reduce erosion.

Pest Management

Organic pesticides are derived from naturally occurring sources. These include living organisms such as the bacteria Bacillus thuringiensis, which is used to control caterpillar pests, or plant derivatives such as pyrethrins (from the dried flower heads of Chrysanthemum cinerariifolium) or neem oil (from the seeds of Azadirachta indica). Mineral based inorganic pesticides such as sulfur and copper are also allowed. In addition to pesticides, organic pest control integrates biological, cultural, and genetic controls to minimize pest damage. Biological control utilizes the natural enemies of pests, such as predatory insects (e.g., ladybugs) or parasitoids (e.g., certain wasps) to attack insect pests. Pest cycles can be disrupted with cultural controls, of which crop rotation is the most widely used. Finally, traditional plant breeding has produced numerous crop varieties that are resistant to specific pests. The use of such varieties and the planting of genetically diverse crops provide genetic control against pests and many plant diseases.

Genetically Modified Organism (GMO)

Genetically modified organism (GMO), organism whose genome has been engineered in the laboratory in order to favor the expression of desired physiological traits or the generation of desired biological products. In conventional livestock production, crop farming, and even pet breeding, it has long been the practice to breed select individuals of a species in order to produce offspring that have desirable traits. In genetic modification, however, recombinant genetic technologies are employed to produce organisms whose genomes have been precisely altered at the molecular level, usually by the inclusion of genes from unrelated species of organisms that code for traits that would not be obtained easily through conventional selective breeding. Genetically modified organisms (GMOs) are produced using scientific methods that include recombinant DNA technology and reproductive cloning. In reproductive cloning, a nucleus is extracted from a cell of the individual to be cloned and is inserted into the enucleated cytoplasm of a host egg (an enucleated egg is an egg cell that has had its own nucleus removed). The process results in the generation of an offspring that is genetically identical to the donor individual. The first animal produced by means of this cloning technique with a nucleus from an adult donor cell (as opposed to a donor embryo) was a sheep named Dolly, born in 1996. Since then, a number of other animals, including pigs, horses, and dogs, have been generated by reproductive cloning technology. Recombinant

DNA technology, on the other hand, involves the insertion of one or more individual genes from an organism of one species into the DNA (deoxyribonucleic acid) of another. Whole-genome replacement, involving the transplantation of one bacterial genome into the cell body, or cytoplasm, of another microorganism, has been reported, although this technology is still limited to basic scientific applications. GMOs produced through genetic technologies have become a part of everyday life, entering into society through agriculture, medicine, research, and environmental management. However, while GMOs have benefited human society in many ways, some disadvantages exist; therefore, the production of GMOs remains a highly controversial topic in many parts of the world.

GMO in Agriculture

Genetically modified (GM) foods were first approved for human consumption in the United States in 1994, and by 2014-15 about 90 percent of the corn, cotton, and soybeans planted in the United States were GM. By the end of 2014, GM crops covered nearly 1.8 million square kilometers (695,000 square miles) of land in more than two dozen countries worldwide. The majority of GM crops were grown in the Americas. Engineered crops can dramatically increase per area crop yields and, in some cases, reduce the use of chemical insecticides. For example, the application of wide-spectrum insecticides declined in many areas growing plants, such as potatoes, cotton, and corn, that were endowed with a gene from the bacterium Bacillus thuringiensis, which produces a natural insecticide called BT toxin. Field studies conducted in India in which BT cotton was compared with non-Bt cotton demonstrated a 30-80 percent increase in yield from the GM crop. This increase was attributed to marked improvement in the GM plants' ability to overcome bollworm infestation, which was otherwise common. Studies of BT cotton production in Arizona, U.S., demonstrated only small gains in yield—about 5 percent—with an estimated cost reduction of $25–$65 (USD) per acre owing to decreased pesticide applications. In China, where farmers first gained access to BT cotton in 1997, the GM crop was initially successful.

Farmers who had planted BT cotton reduced their pesticide use by 50-80 percent and increased their earnings by as much as 36 percent. By 2004, however, farmers who had been growing BT cotton for several years found that the benefits of the crop eroded as populations of secondary insect pests, such as mirids, increased. Farmers once again were forced to spray broad-spectrum pesticides throughout the growing season, such that the average revenue for BT growers was 8 percent lower than that of farmers who grew conventional cotton. Meanwhile, BT resistance had also evolved in field populations of major cotton pests, including both the cotton bollworm (Helicoverpa armigera) and the pink bollworm (Pectinophora gossypiella).

Dryland Organic Farming

Organic farming has assumed immense significance in the dryland areas also. Soil and climatic conditions in India's drylands make them particularly well suited to Organic Farming. These marginal lands, with their marginal soils do not respond well to intensive farming practices. These are actually better suited to low-input farming systems that make ample use of the biodiversity. In turn, organic farming with its central focus on maintaining and improving soil health, its avoidance of pollutants, and its reliance on local inputs and labor, can materially advance the economic and ecological health of the drylands, as well as people who live there. Semiarid and arid dryland soils typically are poor in water-holding capacity as well as organic matter. In some areas, depth of the soil is another limiting factor for agricultural production. Addition of organic matter, a corner stone of organic farming practices, will not only improve the physical condition of these dryland soils, but also greatly improve their ability to supply balanced plant nutrients. In drylands, there is over-exploitation of natural resources mainly because of inappropriate production-enhancing technologies. For example, use of tractor increases wind erosion and damages natural regeneration of trees and grasses. Over-use or improper use of canal irrigation can cause water logging and salinity.

Excessive groundwater pumping has decreased the groundwater table drastically in tube-well irrigated areas. In many locations where intensive-input agriculture systems are followed, soil fertility is decreasing and certain severe pests are becoming resistant to synthetic pesticides. These are all indicators of improper land use, leading to desertification; adoption of organic farming practices suitable for drylands can help to ameliorate these conditions. Due to climatic variability, farming systems in drylands traditionally use a mix of crops, trees, animals, and grasses. Such diversified systems have been found efficient in nutrient recycling and restoration of soil fertility- the basic aims of organic farming; they minimize pest incidence as well. Furthermore, India's traditional farmers possess a rich body of wisdom, based on long observation and practice, concerning soil fertility and pest control management; this can be used to strengthen organic systems. These two factors will also aid the quick development of more efficient, more productive organic farming systems in these areas. In terms of input supply, the drylands are very rich in local resources that are suitable for supporting organic farming.

Soil Fertility in Organic Farming

We have had two decades of large-scale and rapid destruction of fertile agricultural soils in India as a result of the very processes which attempted to increase agricultural productivity. The green revolution paradigm substituted the nutrients cycle with linear flows of purchased inputs of chemical fertilizers and focused on the production of chemical marketable agricultural commodities. Yet, as the Punjab experience has shown, the fertility of soils cannot be restored by NPK from factories, and agricultural productivity necessarily includes returning to the soil a part of the biological products that the soil yields.

Technologies cannot substitute nature and work outside nature's ecological processes without destroying the very basis of production nor can markets provide the only measure of Output and yields. The green revolution technologies created the perception that soil fertility is produced in chemical factories, and agricultural yields are measured only through marketed commodities. Nitrogen fixing crops like pulses were displaced. Millets which have high yields from the perspective of returning organic matter to the soil were rejected as marginal crops. Biological products not sold on the market but used as internal inputs for maintaining soil fertility were totally ignored in the cost-benefit equations of the green revolution miracle. They did not appear in the list of inputs because they were not purchased, and they did not appear as outputs because they were not sold. The FYM has always been one of the principals means of replenishing soil losses. It supplies soil organic matter (SOM) which is an indicator of life, soil health and even its production capacity. Plant biomass is the only input needed for enhancing SOM. Organic manures not only supply nutrients to crops and improve the soil texture in dry lands but also act as mulches. They protect crops against adverse temperature effects, improve seed germination, increase water retention capacity of the soil and create the right micro-climate for the development of beneficial soil microbes. Organically cultivated soils are relatively better attuned to withstand water stress and nutrient loss. Their potential to counter soil degradation is high and several experiments in arid areas have revealed that organic farming may help combat desertification.

In our modern agricultural system, we have forgotten how to feed the soil; we just feed the plants. If we feed the soil, it is necessary to only compensate for the elements that have been exported with the seed. This need can, to some extent, be fulfilled by growing plants like soya bean, which are nitrogen fixing. It is possible in such a manner to develop an organic system with extremely low inputs of fertilizers in the soil. Study conducted in AP on the farmers soil fertility management practices Butterworth and how it helps for the livelihoods of the people, have found that farmers are usually rational decision makers, who weigh the costs of any practice against the potential benefits and attempt to make a net gain. What was 'unproductive 'and waste in the commercial context of the green revolution, is now emerging as productive in the ecological context and as the only route to sustainable agriculture. The solution to the crisis of dying soils cannot lie in the hands of those who created the problem and who looked only at the market, not at the life of the soil. The healing and recovery of soil health will not emerge by continuing to cling to the market as an organizing principle for agriculture. This recovery lies in rediscovering natural ways of renewing and learning, once again, to see that the soil has a right to share its produce to maintain its health. Respecting that right is critical to satisfy our needs. Livestock is an integral part of agriculture and has profound influence on its sustainability. Apart from providing additional income, livestock generates employment in the rural area itself. Livestock contributes directly to agriculture by producing manure and influencing the availability of organic carbon to soil.

Integration of livestock and crop production, or mixed farming, allows the use of animal manure to increase soil fertility. Farmers recognize the benefits of using manure, and with the relatively high costs of mineral fertilizers, manuring could play a greater role in maintaining soil fertility. The livestock component of the farming system is crucial to help maintain soil fertility, supply of draught power and food for the family. The nutrient management system has rather become more closed with the weakened traditional linkages between forest and livestock. Increased income through livestock strengthens the capacity of a household to invest on productivity enhancing measures through purchase of off-farm inputs. Earnings from the landholdings of a majority of marginal, small and semi-medium farmers alone are not adequately sufficient for the household round-the-year and livestock rearing provides an alternative to these smallholders. The livestock economy is changing very fast in Andhra Pradesh. The growth of draught animal stock has slowed down compared to the growth of milch animal stock; the latter is growing relatively fast. Across milch animals, the proportion of cross-breeds is also growing rapidly. The reasons for this include reducing farm size, increasing mechanization, declining area of common property resources (CPRs) and reducing pattern in labor availability. This has important implications for the availability of manure. Local animal breeds being important for livelihoods and sustainable agriculture, should be conserved in situ by strengthening integrated farming and indigenous systems of land-use in which livestock plays a key role in nutrient cycles and the maintenance of soil fertility. A jury of Prajateerpu(farmers) believed that the erosion of livestock biodiversity would increase with the corporate agriculture proposed under Vision 2022'. They specifically called for appropriate training and research as well as for sufficient government support to re-introduce livestock into agriculture. Declining fodder and water resources combined with blanket animal-breeding policies are fueling a downward spiral of loss in livestock genetic diversity, draught power, natural fertilizers, livelihoods and household assets.

Agricultural productivity can be improved by better integrated crop and livestock systems, recycling crop residues, and the careful use of other available nutrients, opined that a farming system that aims to optimize the income and employment potential of the small farm through concurrent attention to crop and animal husbandry and post-harvest technologies, needs to be fostered more widely. No major research programme in agriculture should be started without a fair understanding of the existing farming systems. There is inadequacy of the animal power in rainfed ecosystems. We need to identify critical and timely requirements of the draught power in the production systems besides the extended use of available draught power during less critical periods. Thus, livestock production, being a self- income generating enterprise, reduces the irregularity and uncertainty in income from farm business (Anonymous, undated). A combination of agriculture with dairy and poultry farming fetches the small farmers more average net income than the other enterprises, with the weakening of forest and livestock linkages, the nutrient management system has become closed. Also, reduction in the common property resource areas will lead to a decrease in the availability of nutrients to soil. To sum-up, livestock is crucial not only to help maintain soil fertility, supply of draught power and food for the farm-household but also to increase agricultural productivity. Modern land-use practices highly influence the factors controlling soil health because, while these techniques increase the short-term supply of material goods, over time these practices might undermine many ecosystem services on regional and global scales. Soil disturbances operate at various spatial and temporal scales and mediate soil spatial heterogeneity. For instance, such disturbances may reduce the biomass of dominant organisms or provoke alterations in the physical structure of the soil substrate. Cultivation intensity reduces C content and changes the distribution and stability of soil aggregates, leading to a loss of C-rich macro aggregates and an increase of C-depleted micro aggregates in soils. Moreover, soil aggregation is a major ecosystem process directly impacted, via intensified land-use, by soil disturbances, or indirectly through impacts on biotic and a biotic factor that affect soil aggregates.

3

Organic Farming benefits from Microorganisms

Organic Matter

Numerous scientific studies show that soil organic matter improves soil structure so that it is more resistant to erosion and is easier to till, resulting in lower energy use and less greenhouse gas output. Soils with good SOM levels are more efficient at absorbing rainwater and storing it for plants to use in dry periods. Studies show that organic systems get around 30 percent higher yields in periods of drought than conventional systems due to the increase of SOM and its ability to capture and store water for crops. Worldwide, agriculture is responsible for between 11 and 30 percent of greenhouse gas emissions, depending on the boundaries and methodologies used to determine its emissions. According to the United Nations Environment Programme, the estimates of global greenhouse gas emissions in 2010 were 50.1 gigatons of carbon dioxide equivalent (Gt CO_2e) per year. To keep global mean temperature increases below 2°C compared to pre-industrial levels, GHG emissions will have to be reduced to a median level of 44 Gt CO_2e in 2020. This means that the world will have to reduce the current level of emissions by 6.1 Gt CO_2e by 2020 and reduce it every subsequent year. According to the latest World Meteorological Organization figures, the levels of GHG pollution in the atmosphere and the oceans are the highest in history and are still increasing.

There are different types of microorganisms are used in the bio-fertilizers. Bio-fertilizer being essential components of organic farming play vital role in maintaining long term soil fertility and sustainability. Organic farming has emerged as an important priority area globally in view of the growing demand for safe and healthy food and long-term sustainability and concerns on environmental pollution associated with more use of agrochemicals. Though the use of chemical inputs in agriculture is inevitable to meet the growing demand for food in world, there are opportunities in selected crops and niche areas where organic production can be encouraged to tape the domestic export market. Bio-fertilizer is a substance which contains living microorganisms which, when applied to seed, plant surf aces, or soil, colonizes the rhizosphere or the interior of the plant and promotes growth by increasing the supply or availability of primary nutrients to the host plant areas with the objective of increasing number of such microorganisms and accelerate those microbial processes which augment the availability of nutrients that can be easily assimilated by plants. More use of synthetic fertilizers has led to the pollution and contamination of the soil, which has polluted water basins, destroyed micro-organisms and friendly insects, making the crop more prone to diseases and reduced soil fertility. Demand is much higher than the availability. Depleting fossil fuels and increasing cost of fertilizers. This is becoming unaffordable by small and marginal farmers. Depleting soil fertility due to widening gap between nutrient removal and supplies. Growing concern about environmental hazards. Increasing threat to sustainable agriculture. Nitrogen, phosphorus and potassium are the three major nutrients required for healthy plant growth. Cow urine contains significant amounts of both nitrogen and potassium. Research shows that only 20% of nitrogenous materials consumed by cattle is absorbed and 80% is excreted in urine and dung. 52% of Nitrogen returns in the form of urine while 28% return in form of dung 61-87% phosphorus and 82-92% potash was also obtained from urine.

- After analysis it is found that micronutrients increase in soil after application of cow urine.
- Color of leaves is greener than with the use of urea application.
- Residual effect of cow urine is present in next crop.
- Improves the Soil Texture
- Creates good environment in soil for earthworm growth.
- Cow urine sprayed after 14 days of storage in cool place works as insecticide against aphids and other insects.
- It serves as growth promoter of plants.

The spraying of urine not only provides nitrogen for plants but also protects the plants from aphid and other insects and provides resistance to diseases.

Neem

As opposed to synthetic pesticides, neem-based pesticides have multiple modes of action against insects, and therefore rarely induce resistance. Neem acts on the insect's hormonal system, not on the digestive or nervous systems, which makes it very difficult for future generations of insects to develop any resistance.

Green Chili extract

Green Chilies are a natural source of antioxidants which protects the cells against infections s caused by free radicals. This ensures to protect and prevent against any kind of damage. The free-radical scavengers in green chilies comprises Vitamins A, B and C which are essential to strengthen the overall immunity required to fight against various types of infections and disorders.

Garlic extract

Garlic is a highly nutritive value extract; it contains large number of important enzymes and more than 200 biochemical compounds such as antioxidants and vitamins. It contains high contents of volatile and sulphur compounds such as aliin, allicin, ajoene, allylpropyl, diallyl, trisulfide, sallylcysteine, vinyldithiines, sallylmercaptocystein and others. In addition, garlic is considered as a source of vitamins (especially vitamin B complex and vitamin C), antioxidants, flavonoids and minerals (especially P, K and Se) being even considered a rich source of other non-volatile phytonutrients with important medicinal and therapeutic properties, from which a particular emphasis is given to flavonoids, saponins and sapogenins, phenolic compounds, nitrogen oxides, amides and proteins.

Ascorbic acid (vitamin C) is one of those antioxidants. Ascorbic acid is an abundant molecule in garlic plants. It has also been shown to play multiple roles in plant growth such as in cell division, cell wall expansion and other developmental processes. It also acts as an antioxidant as it detoxifies H_2O_2 which is formed by the dismutation of O–2. Therefore, the objective of this study was to compare the effect some foliar application with garlic extract, ascorbic acid and nicotinamide on growth, photosynthetic pigments content, and yield and yield components.

Tobacco Bio-oil

Tobacco Bio-oil Kills Agricultural Pests Cigarette smoking continues to be the leading cause of preventable death and disease in the United States, but tobacco has potentially beneficial uses as well as deadly ones. Gardeners have long known that homemade mixtures of tobacco and water can kill insect pests. But these homemade brews kill desirable insects, too, and could poison animals that ingest them. Heated finely ground tobacco leaves to 500°C in a vacuum; a process called pyrolysis, and then collected the condensate. The bio-oil was tested against the Colorado potato beetle (Leptinotarsa decemlineata), 11 fungi, and 4 bacteria, all of which are agricultural pests. The bio-oil blocked the growth of the bacteria Streptomyces scabies and Clavibacter michiganensis and the fungus Pythium ultimum. S. scabies causes a common potato scab disease that makes potatoes unmarketable, C. michiganensis kills young plants and deforms fruits, especially tomatoes, and P. ultimum kills seedlings of eggplant, peppers, lettuce, chilies, tomatoes, and cucumbers. The bio-oil also killed 100% of Colorado potato beetles, a resistant pest that can destroy potato crops. Nicotine, a key toxin in tobacco, has known insecticidal properties on its own. But even after removing nicotine from the bio-oil, it still potently killed these few pests.

Dhatura

Datura also known as **Thorn Apple**, though considered as one of the deadliest plant species, for its super toxic components, when consumed raw is also surprisingly a powerhouse of medical components if purified properly. Be it the leaves, fruits, flowers, stem or roots, Datura has been traditionally used in both folklore medications and alternative therapies. However, due to the strong hallucinogenic properties of the plant, Datura is often used to relieve asthmatic symptoms and reduce the pain during the surgery and bone setting procedures. Though a strong narcotic plant, datura offers umpteen health benefits and agricultural benefits, it helps to improving fertility of soil.

Calotropis (Aak) extract

The plant can help improve soil water conditions and also acts as a soil binder. Plants often escape from cultivation, especially when growing on sandy soils. It often self-sows freely on overgrazed land and has been used as an indicator of exhausted soil.

Ginger extract

Ginger (Zingiber officinale Roscoe) is a common and widely used spice. It is rich in various chemical constituents, including phenolic compounds, terpenes, polysaccharides, lipids, organic acids, and raw fibers. The health benefits of ginger are mainly attributed to its phenolic compounds, such as gingerols and shogaols. Accumulated investigations have demonstrated that ginger possesses multiple biological activities, including antioxidant, anti-inflammatory, antimicrobial, anticancer, neuroprotective, cardiovascular protective, respiratory protective, anti- obesity, antidiabetic, antinausea, and antiemetic activities, bioactive compounds and bioactivities of ginger, and the mechanisms of action are also discussed.

Biofertilizers

A live formulation of micro-organisms (bacteria, fungi and algae) is used for fertilization of farm lands in the aspect of fixation of N_2, solubilization and mobilization of phosphorus, increasing organic carbon content, balanced C/N ratio, plant growth promotion by increasing nutrient absorption, antagonistic activity against plant pathogens, production of hormones etc. that are beneficial for agriculture. The Gazette of India (2006) defined bio-fertilizer as a product containing carrier based (solid or liquid) living micro-organisms that are agriculturally useful in terms of nitrogen fixation, phosphorous solubilization or nutrient mobilization. Bio-fertilizers were promoted through integrated plant nutrient systems (IPNS) that involved combining fertilizers, organic/green manures and bio-fertilizers to sustain crop production, maintaining soil productivity, health and diversity.

As a safe alternative to chemical fertilizer, it minimizes the ecological disturbance and is cost effective, eco-friendly and when they are required in bulk can be generated at the farm itself. They increase crop yield up to 10-40% and fix nitrogen up to 40-50 Kg and parental inoculums are sufficient for growth and multiplication of organisms in subsequent years. They improve soil texture, pH, WHC and other properties of soil. They produce plant growth promoting substances IAA, amino acids, vitamins etc. Bio-fertilizers contained 3.5% - 4% nitrogen, 2% - 2.5% phosphorus and 1.5% potassium. In terms of N: P: K, it was found to be superior to farmyard manure and other type of manure. are effective in inducing plant growth, enhance seed germination and root and shoot growth, decomposition of organic materials and enrichment of compost.

Microorganisms Used as Bio-Fertilizers

Nitrogen Fixing Bacteria

Rhizobia: Nitrogen, as a principal component of cell bimolecular (Amino acid, proteins, enzymes), all plants need it as a sole source for their metabolism. Mostly leguminous plants require high amount due to their protein demand and can't fix atmospheric N2 due to high energy expenditure. So, Legumes plants have root nodules, where atmospheric nitrogen fixation is done by bacteria belonging to genera, Rhizobium, Bradyrhizobium, Sinorhizobium, Azorhizobium and Mesorhizobium collectively called as rhizobia, belong to á-Proteobacteria. When rhizobial culture is inoculated in field, pulse crops yield can be increased due to rhizobial symbiosis. Rhizobium can fix 15-20 N/ha and increase crop yield up to 20% in pulses.

Azotobacteracae (e.g., Azotobacter): They are the free living, non-symbiotic, aerobic, photoautotrophic bacteria. They secrete vitamins, gibberellins, naphthalene, acetic acid and other substances that inhibit certain root pathogens and improve root growth and uptake of plant nutrients. It occurs in the roots of Paspalumnotatum (tropical grasses) and other spp. and adds 15- 93 Kg N/ha/annum on P. notatum roots. Azotobacter indicum occurs in acidic soil in sugarcane plant roots. It can apply in cereals, millets, vegetables and flowers through seed, seedlings soil treatment.

Spirillaceae (e.g., Azospirillum and Herbaspirillum): These are gram negative, free living, associative symbiotic and non-nodule forming, aerobic bacteria, occurs in the roots of dicots and monocot plants i.e., corn, sorghum, wheat etc. It is easy to culture and identify. Azospirillum is found to be very effective in increasing 10-15% yield of cereal crops and fixes N2 upto 20-40% Kg/ha. Different A. brasiliense strains inoculation in the wheat seed causes increase in seed germination, plant growth, plumule and radicle length. Herbaspirillum species occurs in roots, stems and leaves of sugarcane and rice. They produce growth promoters (IAA, Gibberillins, Cytokinins) and enhance root development and uptake of plant nutrients (N, P & K).

Acetobacter diazotrophicus: Another diazotroph is Acetobacter diazotrophicus occurs in roots, stem and leaves of sugarcane and sugar beet crops as nitrogen fixer and applied through soil treatment. It also produces growth promoters e.g., IAA and helps in nutrients uptake, seed germination, and root growth. This bacterium fixes nitrogen up to 15kg/ha/year and enhance up to 0.5-1% crop yield.

Cyanobacteria (Blue green algae): Nostoc, Anabaena, Oscillatoria, Aulosira, Lyngbya etc. are the prokaryotic organisms and phototropic in nature. They play an important role in enriching paddy field soil by fixing atmospheric nitrogen and supply vitamin B complex and growth promoting substance which makes the plant grow vigorously. Cyanobacteria fixes 20-30 Kg/N/ha and increase 10-15% crop yield when applied at 10 Kg/ha. Both free-living as well as symbiotic cyanobacteria (blue green algae) used in rice cultivation in India.

Azolla-Anabaena symbiosis: It is a free floating, aquatic fern found on water surface having a cyanobacterial symbiont Anabaena azolla in their leaves. It fixes atmospheric nitrogen in paddy field and excrete organic nitrogen in water during its growth and also immediately upon trampling. Azolla contributes nitrogen, phosphorus (15-20 Kg/ha/month), potassium (20-25 kg/ha/month) and organic carbon etc. and increases 10-20% yield of paddy crops and also suppresses weed growth. Azolla also absorbs traces of potassium from irrigation water and can be used as green manure before rice planting. Azolla spp. are metal tolerant hence, can be applied near heavy metal polluted areas.

Phosphate Solubilizing Bacteria: Solubilization of inorganic tricalcium and rock phosphate occurs by most of bacteria belong to genus: Pseudomonas, Bacillus, Acrobacter, Nitrobacter, Escherichia, Serratia spp. and specially Pseudomonas striata, Bacillus polymyxa are the bacteria have phosphate solubilizing ability. Phosphobacterin' are the bacterial fertilizers containing cells of Bacillus megatherium var. phosphaticum, prepared firstly by USSR scientists. They increased about 10 to 20% crop yield and also produces plant growth promoting hormones which helps in phosphate solubilizing activity of soil.

Phosphate Solubilizing Fungi: Some fungi also have phosphate dissolving ability e.g., Aspergillus niger, Aspergillus awamori, Penicillium digitatum etc. All the microbes solubilize phosphate by production of organic acids.

Plant Growth Promoting Rhizobacteria (PGPR)

The group of bacteria that colonize roots or rhizosphere soil and beneficial to crops are referred to as plant growth promoting rhizobacteria (PGPR). They are also called as microbial pesticides e.g., Bacillus spp. and Pseudomonas fluorescence, Serratia spp. and Ochrobactrum spp. are able to promote growth of plants. The PGPR inoculants currently commercialized that seem to promote growth through at least one mechanism; suppression of plant disease (termed Bio- protectants), improved nutrient acquisition (termed Bio-fertilizers), or phytohormone production (termed Bio-stimulants). Species of Pseudomonas and Bacillus can produce as yet not well characterized phytohormones or growth regulators that cause crops to have greater amounts of fine roots which have the effect of increasing the absorptive surface of plant roots for uptake of water and nutrients. These PGPR are referred to as Bio-stimulants and the phytohormones they produce include indole-acetic acid, cytokinins, gibberellins and inhibitors of ethylene production.

Mycorrhiza: Mycorrhiza is symbiosis between some specific root inhabiting fungi and plant roots and used as Bio-fertilizers. They absorb nutrients such as manganese, phosphorus, iron, sulphur, zinc etc. from the soil and pass it to the plant. Mycorrhizal fungus increases the yield of crops by 30-40% and also produces plant growth promoting substances.

Advantages of Biofertilizer

- Suitable for all crops and increases the yield.
- Earthworm count increases in the soil. Earthworm helps to keep soil quality porous which has a higher water holding capacity.
- Acts as an agent to increase the microbial count & friendly bacteria in the soil.
- Improves pH of the soil.
- This biofertilizer can be made within 4-5 days so it can be used effectively and frequently.
- Reduces costs of chemical fertilizers.
- Improves aeration, bring up minerals from deep in the field.

Benefits of Stable Soil Organic Matter

There are numerous benefits to having a relatively high stable organic matter level in an agricultural soil. These benefits can be grouped into three categories:

Physical Benefits

- Enhances aggregate stability, improving water infiltration and soil aeration, reducing runoff.
- Improves water holding capacity.
- Reduces the stickiness of clay soils making them easier to till.
- Reduces surface crusting, facilitating seedbed preparation.

Chemical Benefits

- Increases the soil's CEC or its ability to hold onto and supply over time essential nutrients such as calcium, magnesium and potassium.
- Improves the ability of a soil to resist pH change; this is also known as buffering capacity.
- Accelerates decomposition of soil minerals over time, making the nutrients in the minerals available for plant uptake.

Biological Benefits

- Provides food for the living organisms in the soil.
- Enhances soil microbial biodiversity and activity which can help in the suppression of diseases and pests.
- Enhances pore space through the actions of soil microorganisms.

Condition Affecting Soil Microorganism

Many conditions of the soil effect soil microorganisms. Among the most important are the supplies of oxygen, moisture, temperature, nature of the soil, organic matter and pH.

Soil Nutrients

The Indian soils are suffering from degradation due to improper management and faulty land use practices. The use of plant nutrients per hectare is relatively low and imbalanced, and this is one of the major causes for soil nutrient deficiencies and low crop yields in India. Hence, it is essential to protect the soil from further degradation to produce healthy crops. Soil analysis results provide valuable information about the distribution pattern of plant nutrients in Indian soils, calcium and magnesium contents as well as action exchange capacity (CEC) of these soils are low and about one-third of the Indian soils are acidic. Sulphur deficiency is also seen in the soils due to use of sulphur free fertilizers and intensive use of soil. It was found that the micronutrients availability in acidic soils is relatively high. However, the deficiency of these nutrients has also been noticed in the lands under intensive cultivation.

It was observed that after 1960, micronutrients deficiencies; especially of Zn and Fe was noticed in light textured soils of India. These deficiencies might be due to continuous use of chemical fertilizers under intensified cropping and ignorance of organic manures. Later on, multiple nutrient deficiencies were observed in soil for N, NP, NPFe, NPFeZn, NPFeZnK, NPZnKS, NPZnKSB and NPZnKSMnMo within a period of 1960 to 2005. Multinutrient deficiencies are emerging for Zn + Fe in swell-shrink soils, Zn + Mn or Zn + Fe + Mn in alluvial soils of Indo-Gangatic plains, Zn + Fe, Zn + B, Zn + Fe + B in highly calcareous soils of Bihar, Gujarat, Zn + B in acid leached Alfisols, red and Lateritic soils of India. Despite application of adequate quantity of NPK, the yield remains low due to hidden deficiencies of micronutrients like Zn, Fe, Cu, Mn and B.

Micronutrient deficiency in Indian soils has emerged as one of the major constraints to crop productivity. Zinc, iron, boron and manganese deficient areas are relatively more, even the deficiencies of copper and molybdenum are observed in many districts. These deficiencies are mainly due to soil acidity in eastern and southern states and soil alkalinity in northwestern states. Crops grown on such soils encounter nutritional disorders and toxicities. National Food Security Mission (NFSM), a strategic initiative of Ministry of Agriculture, Govt. of India, has endeavored to address these constraints by creating awareness among farmers and promoting use of soil ameliorants and micronutrients in major states of the country. Under such situation, maintenance of soil fertility becomes the foremost task before the farmers of India. A fertilizer being a costly input, but it is essential and it will remain an unavoidable input in agricultural production. Hence, it is necessary to know all aspects of fertilizer application. Fertilizers should be used judiciously based on the fertility status of the soil and actual need of the plants. The fertilizer recommendation for a crop in particular region may sharply differ from that in a different region for the same crop. Scientific observations have shown that there is no relation between the total nutrient content of the soil and availability of nutrients to the plant roots. Total nutrient content of a soil may be high, but the availability is very low. Therefore, using total soil composition as a guiding criterion for estimating the availability of nutrient status is incorrect. For example, in some laterite soils total contents of molybdenum, zinc, manganese, copper and boron vary from 1-18 ppm, 2-1600 ppm, 37-4600 ppm, 2-1000 ppm and 6-360 ppm, respectively; while available nutrient content at a given time is found only up to 2% of total copper, 0.5 % of total boron, 1-5 % of total zinc and less than 10% of total molybdenum.

The nutrients, which remain dissolved in the soil solution, are available to the plant root for absorbance. Nutrients absorbed on the soil exchange complex like clay and humus, and fixed with in organic compounds; refill soil solution when the plant roots absorb nutrients from soil solution. However, nutrients, which are immobilized in humus or microbial body, can be availed by the crop roots after decomposition of humus and dead microbial cells. Availability of nutrients in the soil varies depending on a number of factors. For example, under moisture stress during draught conditions, movement of nutrients to roots may be restricted even though the element is present in the soil in available form. Soil reaction also affects available nutrient status of the soil. The well-known effect in acid soil is the too high availability of a few and too low availability of remaining plant nutrients. Some amendments and management practices can help correct extreme pH values and bring the soil to a favorable level of reaction. Millions of acres of arable land worldwide, particularly in arid and semi-arid regions, are deficient in plant-available micronutrients and this can markedly affect human nutrition. Major reason for the widespread occurrence of deficiency of micronutrients is the low availability of micronutrients to plant roots rather than their low concentration in soils. Low solubility of most micronutrient cations like copper, iron, manganese and zinc in soils means that after the addition to alkaline soil as the soluble form, the metal is rapidly sorbet or precipitated. Micronutrient deficiencies, particularly Zn and B are one of the emerging constraints in sustainable crop production even in rainfed areas.

Soil pH and Soil Minerals

Many mineral forms are insoluble compounds, which are not available to the plant for absorption until they are converted into free mineral ions. This occurs when they are exposed to the low pH of the soil. However, if acid output is low, the minerals may not be solubilized and thus cannot be absorbed. This is only the first of many obstacles that minerals encounter during the plant growth. Micronutrients normally have an electrical charge on them, e.g., calcium and magnesium are both $^{+}2$ charge, whereas soil particles are negative in charge. Under this condition calcium and magnesium reacts with the soil and become insoluble and are not available for plants to use. In addition, calcium and magnesium compete with each other for entry into the plant since they have the same charge. Zinc has important functions in the synthesis of auxin or indoleacetic acid (IAA) from tryptophan. Due to alkaline and calcareous nature, soils are more prone to zinc deficiency. Zinc solubility is highly soil pH dependent and decreases 100-fold for each unit increase in pH, and pH is more important than any other single property for controlling Zn mobility in soils. The low availability of Zn in high pH calcareous soils is due to the adsorption of Zn on clay or $CaCO_3$. In addition, high concentrations of HCO_3-inhibit Zn uptake and translocation.

Boron is the only micronutrient to exist in solution as a no ionized molecule over soil pH ranges suitable for the growth of most plants. Increasing soil pH decreases B availability by increasing B adsorption onto clay, Al and Fe hydroxyl surfaces, especially at high soil p^H. Boron deficiency is common for plants grown in arid, semiarid, and heavy rainfall areas in calcareous, sandy, light textured, acid, and low organic matter soils. The factors affecting B uptake include soil type, B content, soil pH, amount of water soil receives, and plant species. Boron is very vulnerable to leaching so its deficiency can temporarily be expected during and after monsoon rains especially on coarse textured soils. Boron is taken up from the soil only at higher soil temperatures than are other elements. The role of iron in photosynthesis, nitrite and sulfate reduction and N_2 assimilation is well established. Although it is the fourth most abundant element in soils, yet its deficiency is wide spread and is by far the most difficult to correct especially in calcareous soils. The solubility of Fe decreases by~1000-fold for each unit increase of soil pH in the range of 4 to 9 compared to ~100-fold decreases in the activity of Mn, Cu, and Zn. Minimum Fe solubility occurs between pH 7.5 and 8.5, which is the pH range of many calcareous soils. The reactions of metal nutrients in soil can be avoided by making the chelates of these metal nutrients.

Chelating agents are organic compounds that bind the metal and increase their water solubility. Common chelating is ethylene dimine tetra acetic acid (EDTA) and dimethyl amine penta acetic acid (DTPA) and these molecules increase micronutrient solubility through reversal of charge on the metal. The metallic cation M^{2+} becomes ML^{n-} where M is the micronutrient cation and L the chelator, i.e., the chelate makes the micronutrient anionic. It is well documented that both EDTA and DTPA obviously increase the solubility of micronutrient cations in soil and help their diffusion to plant roots. However, the high zinc deficiency.mZinc solubility is highly soil pH dependent and decreases 100-fold for each unit increase in pH, and pH is more important than any other single property for controlling Zn mobility in soils. The low availability of Zn in high pH calcareous soils is due to the adsorption of Zn on clay or $CaCO_3$. In addition, high concentrations of HCO_3-inhibit Zn uptake and translocation. Increasing soil pH decreases B availability by increasing B adsorption onto clay, Al and Fe hydroxyl surfaces, especially at high soil pH. Boron deficiency is common for plants grown in arid, semiarid, and heavy rainfall areas in calcareous, sandy, light textured, acid, and low organic matter soils.

Microorganisms are Beneficial to Crops

The groups of microorganisms that colonize roots or rhizosphere in the crop roots are referred to as plant growth promoting rhizobacteria (PGPR). These microorganisms are beneficial to crops. PGPR referred to as bio stimulates and the phytohormones. They produce indole-acetic acid, cytokinins, gibberellins and inhibitors of ethylene production. During late 1978, Kloepper was the first to recognize plant growth promoting rhizobacteria who told that these are major player in crop production in temperate annual crop production. Thus, science of PGPR is relatively young in competition to nitrogen fixing bacteria. Plant growth promoting rhizobacteria (PGPR) can stimulate growth by one or more different mechanisms. The direct mechanisms include production of growth hormones like IAA, phosphate-solubilization and uptake of iron, whereas indirect mechanisms include check on phytopathogens by the release of HCN, antibiotics and siderophores. Among PGPR, fluorescent pseudomonads (FLPs) possess several properties best suited for survival and colonization in the rhizosphere environment. Modern agriculture has become more productive, sustainable, and environmentally friendly. Therefore, the use of rhizobacterial in agriculture for sustainable crop production is becoming more and more important. The present study was aimed to determine the effect of Pseudomonas fluorescens bacterium on the growth of plant, yield and seed nutrient content in lentil. Lentils in production and food systems in West Asia and Africa. International center for agricultural research in the dry areas (ICARDA), Aleppo, Syria. Grain Legumes 57:46-48. The field experiments were carried out during two seasons of 2013-2014 and 2014-2015 under the conditions of Southeastern Anatolia of Turkey. The experiment consisted of four Pseudomonas fluorescens (no treatment; inoculated seeds; foliar treatment and inoculated seeds + foliar treatment) applications. The results showed that the application of Pseudomonas fluorescent rhizobacteria had significant effects on plant growth, yield, and yield components and seed nutrient content in lentil. The results also showed that the inoculated seed + foliar application are more important for yield and yield components, and foliar application is more important for seed nutrient content.

Consequently, studies on the use of such beneficial microorganisms are necessary for a sustainable ecological agricultural and production. Plants require approximately 30 μmol l-1 of phosphorus for maximum productivity, but only about 1 μmol l-1 is available in many soils. Therefore, the unavailability of phosphorus in many soils has been recognized as a major growth limiting factor in agricultural and horticultural systems. This necessitates the application of soluble forms of phosphorus in the form of phosphate fertilizers, which in itself has constraints in that it too is rapidly immobilized (fixed) to insoluble forms upon its application in the soil due to its reaction with aluminum and iron minerals. The efficiency of applied phosphorus rarely exceeds 30% due to fixation in soil. It is also lost as a result of run-off and leaching, leaving as little as 10-20% available for plant utilization. Phosphate fertilizers are dependent on phosphorus derived from phosphate rock, which is a non-renewable resource and current global reserves may be depleted in 50-100 years. Therefore, exploring alternative forms of agriculture, where nutrient conservation is key, is of vital importance.

Since 1990s some researchers have reported the potential of PGPB to promote plant growth and enhance the yield of crops in different soil and environment; in particular, Azospirillum, Azotobacter, Bacillus and Pseudomonas spp. were studied and applied in herbaceous crop systems as biofertilizers. Some promising and controversial results on the mechanisms PGPB-plant root were reported. There is not an official definition of biofertilizers; therefore, we refer to the definition of biofertilizer is a substance containing living microorganisms which, when applied to seed, plant surfaces, or soil, colonizes the rhizosphere or the interior of the plant and promotes growth by increasing the supply or availability of primary nutrients to the host plant. This definition separates biofertilizers from organic fertilizers, which contain organic compounds and directly increase soil fertility. Therefore, biofertilizers should contain living organisms, which improve nutrient use efficiency of the host plant, through different mechanisms. The production of growth substances such as exogenous microbial auxins and gibberellins that showed stimulatory effect on plant growth. In the current study, the rhizobacterium Pseudomonas fluorescens R68 (PFR68) isolated from Western Ghat was analyzed for its growth enhancement effect on the leafy vegetable Amaranthus tricolor (L.). One month of field growth of PFR68 inoculated A. tricolor has found to have enhanced growth parameters such as leaf number (1.57fold), root number (1.76- fold), shoot length (1.28-fold) and fresh weight (2.31-fold). The treatment also improved soil fertility in terms of Nitrogen, Phosphorus and Potassium content. In addition to this, application of PFR68 along with 50% NPK augmented the available Nitrogen and Phosphorus content in soil. In conclusion, field performance of PFR68 on growth of A. tricolor confirms its promises to develop into plant probiotic formulation.

PGPR in Agriculture

Due to the increase in both human population growth and environmental pressure, it is necessary to raise agricultural productivity without enhancing environmental footprint, with in this context; soil inoculation with PGPB (Plant Growth Promoting Bacteria) may be considered a promising tool of integrated management systems. In particular, PGPB may improve plant growth either directly, by facilitating resource use or modulating plant hormone levels, or indirectly by decreasing the inhibitory effects of various pathogenic agents. PGPB comprise different functional and taxonomic groups of bacteria like Pseudomonas, Bacillus, Rhizobium and others. Their ability to either mobilize mineral or organic bound nutrients from the pedosphere or to fix atmospheric N2 and make it available to the plants is a crucial feature in their application. It is estimated that by 2050 the global population reaches to ~9 billion which demands ~60% extra food from the present, which creates extra pressure on the farmers to achieve the need of food. The farmers currently utilized huge number of chemical fertilizers and pesticides to enhance the food production and disease management.

Activities of PGPR

Pseudomonas fluorescens, a PGPR enhances the plant growth productivity by production of metabolites such as siderophores, HCN, chitin, and beta-1, 3-gluconase, antibiotic production which control the pathogenic microorganism in rhizosphere. PF produces indole acetic acid, phosphate solubility and vitamin by which plant root elongates and rhizosphere bacterial population is also improved. These chemical fertilizers not only affect the texture and productivity of soil but also the health of plants, humans, and environment. From the last two decades, plant growth-promoting bacteria (PGPB) is one of the best choices as plant and soil inoculants to enhance the plant growth and disease management, due to their less impact on environment and eco-friendly nature. In this chapter, we focused on one of the most broadly used bacterial genus, Pseudomonas, which is present in the rhizosphere as well as within the tissue of the plants and used in sustainable agriculture for the enhancement of growth promotion and disease management. In literature some data are available on the use of commercial PGPB, while less efforts have been made on the study of the effect of autochthonous PGPB isolated from soils on sustainability of cropping systems; thus, a literature survey on these aspects was carried out with special focus on wheat, a staple food for a large part of world population. In particular, the main topic of this review is the potential of PGPB to enhance use efficiency of agro-environmental resources focusing on the interaction PGPB-wheat for improving nitrogen use efficiency.

HCN Production

Pseudomonads produce HCN which control the root rot pathogen. Pseudomonas extort their beneficial effect on plant, has also reported that Pseudomonas strains produced HCN. Chitin, and beta-1, 3-gluconase, antibiotic production: Hyper parasitism and lysis are the most important form of biological control of soil born plant pathogens by microbial antagonist. The mechanism appears to involve the enzymatic hydrolysis of chitin and glucan component of fungal hyphal wall. The lytic activity of bacterial and fungal antagonist has been reported mainly due to the extra cellular lytic enzyme chitinase and beta-1, 3- gluconase. These enzymes have been found to be responsible for the control activity of some bacteria.

Siderophore Production

Siderophores are low molecular weight compounds that are produced under iron-limiting conditions, chelate the ferric ion (Fe 3+) with a high specific activity, and serve as vehicles for the transport of Fe (III) into a microbial cell. In rhizosphere, the availability of iron for microbial assimilation is extremely limiting. To survive in such environment, organisms secrete iron binding ligands called siderophores which can bind the ferric ion and make it available to the host organisms. Recent research as have shown the involvement of these compounds in plant growth promotion. The root colonizing bacteria, which directly or indirectly stimulate plant growth, are referred as plant growth promoting rhizobacteria (PGPR). They promote growth of several annual crops by increased uptake of nitrogen and phosphorus, synthesis of plant growth promoting substances and uptake of iron through siderophores. PGPR produce siderophores having greater binding potentials. The production of siderophores can be classified as indirect mechanism when they are employed as means of biocontrol by microorganisms that do not use any other mechanism. By sequestering iron, PGPBs siderophore-producing reduce the availability of this element necessary for the growth of pathogens. Siderophore mediated iron scavenging in Gram-negative transport is better studied PGPB than Gram-positive PGPB.

Antibiotic Production

Antibiotic production by microorganisms is one of their more interesting features, particularly from a medical and commercial point of view. More than 10,000 antibiotics and similar bioactive metabolites have been isolated from microbes, with approximately 500 new classes of low-molecular-weight compounds published every year. In monetary terms, antibiotics currently are the most important group of microbial biotechnological products, with an estimated world sale in excess of $15 billion. The primary classes are cephalosporins, penicillin, and tetracyclines, and the majority of these agents are produced by Streptomyces (and other actinomycetes) and various Bacillus species. Their primary use is in the treatment of human infectious diseases, although a significant number have agricultural and veterinary applications. Antibiotics are made by secondary metabolic pathways that use common metabolites in less specific and, sometimes, more intricate ways than primary metabolism. The polyketide antibiotics, for example, are made from simple fatty acids by a pathway that superficially resembles the one used to make long-chain fatty acids, but the resulting compounds exhibit a range of structural complexity far surpassing the simple hydrocarbon framework of the essential fatty acids.

Recently, it has become apparent that yields of secondary metabolites, including antibiotics, can also be improved by overcoming rate-controlling biosynthetic steps through genetic techniques. In addition, metabolic engineering techniques are applied in order to modify known antibiotics to improve their properties and also to synthesize new forms of antibiotics. For years antibiotic production in filamentous fungi and Streptomyces was improved by random mutation/screening and, to a lesser extent, by selecting mutants that over-produced primary metabolic precursors of antibiotics. Work spanning four decades developed production strains less amenable to improvement by these traditional techniques. The application of recombinant DNA technology was based on the development of genetic transformation systems for β-lactam-producing organisms and cloning of biosynthetic genes. The ability to transform industrially important organisms (such as P. chrysogenum and C. acremonium) provided a powerful tool for the precise manipulation of biosynthetic pathways and an avenue for practical applications, such as gene dosage studies for possible limiting steps or gene disruptions to alter final products. The discovery that many antibiotics genes are clustered, and also those certain genes of related pathways exhibit cross-hybridization, has opened new avenues in this area. Gene clustering facilitates cloning, and the fact that these genes are often positively regulated increases the possibility of improving production through overexpression of the genetic regulatory molecule. Overexpression of such regulatory genes caused, for instance, the overproduction of streptomycin, and actinorhodin in wild-type strains.

Phosphate Solubilization

To study the effect of phosphate solubilization microorganisms (PSM) and plant growth promoting rhizobacteria (PGPR) on yield and yield components of corn Zea mays an experiment was conducted at research farm of Sari Agricultural Sciences and Natural Resources University, Iran during 2007. Experiment laid out as split plot based on randomized complete block design with three replications. Three levels of manures (consisted of 20 Mg. ha-1 farmyard manure, 15 Mg. ha-1 green manure and check or without any manures) as main plots and eight levels of biofertilizers (consisted of 1-NPK or conventional fertilizer application. Results showed that farmyard manure application increased row number, ear weight, grain number per year, grain yield, biological yield and harvest index compared to check. According to results in all fertilizer treatments application of PSM and PGPR together could reduce P application by 50% without any significant reduction of grain yield. However, this treatment could not compensate 50% reduction of N application. Keywords-Biofertilizers, corn, PSM, PGPR, grain yield. Large quantities of chemical fertilizers are used to replenish soil N and P, resulting in high costs and severe environmental contamination. Most of phosphorus in insoluble compounds are unavailable to plant.

N2-fixing and P-solubilizing bacteria may be important for plant nutrition by increasing N and P uptake by the plants, and playing a significant role as plant growth promoting rhizobacteria (PGPR) in the biofertilization of crops. Plant autors are with Department of Agronomy and Plant Breeding, Sari Agricultural Sciences and Natural Resources University, Sari, Iran. Growth- promoting rhizohacteria (PGPR) are able exit a beneficial upon plant growth. Nitrogen fixation and P. solubilization production of antibiotic and increased rood dry weight is the principal mechanism for the PGPR. A number of different bacteria promote plant growth, including Azotobacter sp., Azospirillum sp., Pseudomones sp., Bacillus sp. Acetobacter sp. Economic and environmental benefits can include increased income from high yields, reduced fertilizer costs and reduced emission of the greenhouse gas, N_2O as well as reduced leaching of NO_3, N to ground water. Plant growth promoting bacteria are important in managing plant growth because of their effects on soil conditions, nutrient availability, growth and yields.

Indole Acetic Acid

Indole Acetic Acid Indole-3-acetic acid (IAA) is the most common and studied auxin, which plays an important role in cell division, differentiation, and germination of seed and tubers, controls vegetative growth, and initiates lateral and adventitious root formation. Florescence affects photosynthesis, pigment formation, and biosynthesis of various metabolites in stress conditions. The response of IAA varies from species to species in the plant. Some of the plant species are sensitive to the IAA; it varies with the genotypes of the plants. In plant root, endogenous IAA may be suboptimal or optimal for growth.

Interaction of Pseudomonas with another Microorganism

Rhizosphere and their Interaction with Microbes Plant growth in agricultural soils is influenced by many biotic and abiotic factors. Bacteria are the most abundant group of microorganisms coexisting with fungi, protozoa, and algae in the rhizosphere. Since bacteria are the most abundant microorganisms in the rhizosphere, it is highly probable that they influence the physiology and competitiveness in root colonization in plants to a greater extent. These effective colonizing microorganisms are classified according to the interactions with plants; these may be positive, negative, or neutral. Rhizobacterial species inhabit of plant roots, exert positive effect ranging from direct influence mechanisms to an indirect effect during the plant growth and development. Recently the number of PGPR isolates has been used as plant and soil inoculants for the growth and disease management. These isolates belong to all the major families; some of the important species are Pseudomonas, Azospirillum, Azotobacter, Klebsiella, Enterobacter,

Alcaligenes, Arthrobacter, Burkholderia, Bacillus, and Serratia which are broadly used as plant growth-promoting and biocontrol agents. Interaction of Pseudomonas with other Microbe in Rhizosphere is the most prominent habitat for the diverse group of microbes. These microbes interacted with each other and have significant effect on the host plants. Pseudomonas is the most effective root colonizer present in both rhizospheres as well as within the tissue of plants. In the rhizosphere, microbes can have both beneficial and harmful effect on the plants. Plant species get benefits like phytohormone production, availability of nutrients, and abiotic tolerance through the mutualistic interaction with the microorganism in the rhizosphere. These benefits can be provided by different groups of microorganisms. Many studies examined the effect of co-inoculation with different groups of microbes in the soil or plants. The co-inoculation of microbial species in the plant or soil may be Pseudomonas + fungi, Pseudomonas + other bacteria like Bacillus and Azotobacter, or Pseudomonas + different species of Pseudomonas. These multiple agents enhanced the possibility to better response in growth or biocontrol action in the plants due to chance of better survivability of either of the strains in the applied environmental condition. It is also revealed from the previous study that co-inoculation of different microbial species is more effective than single inoculation for biocontrol, combined application of P. fluorescens and Bacillus sp. more effectively controls the Fusarium disease. Co-inoculation of P. fluorescens and Azospirillum stimulated the root growth in spring wheat. The most common inoculation of fungi with Pseudomonas is the mycorrhizal fungi, used to improve the plant growth and nutrition combined use of P. fluorescens with Beauveria bassiana more effective than single application. In case of interaction of Pseudomonas with other PGPR, Bacillus is the most common species. Single or combined application of P. fluorescens with Bacillus is equally effective against bacterial and fungal pathogen. It is concluded from the previous study that combined introductions of Pseudomonas and other PGPR have been more beneficial in growth promotion and in biological control.

Importance of Microorganisms in Organic Farming

While conventional agriculture relies on synthetic fertilizer, chemical pesticides or other technological inputs for crop production, Organic Farming aims to make optimal use of the natural capital of the soil and its microbial population through methods such as the selection of indigenous crop varieties, and the production of crops appropriate to soil conditions. A number of different soil microorganisms are involved in these processes. Many shows potential biocontrol activities against weeds, crop diseases and pests, while Rhizobacteria and mycorrhizal species can play an important role in sustainable fertility management. Microorganisms are currently being used as a replacement for synthetic pesticides and fertilizers for many different crops. A more holistic concept of farming in terms of agroecology has begun to challenge the traditional reductionism approach to the study of agriculture. As a result, we have become more aware of the importance of microorganisms in such processes as soil formation, plant nutrition, and the suppression of plant pathogens, pests and weeds. Moreover, it has become clear that many farming practices- intensive tillage, pesticide use, fertilizer use and monocropping in particular- are directly or indirectly harmful to soil microbes and therefore to the processes mediated by them. It is concern that since much industrialized agriculture now relies heavily on manufactured inputs rather than using natural resources, the sustainability of agriculture is threatened and a move towards resource conserving agriculture is essential. Farming practices must be designed to optimize soil microbial life as part of a wider strategy aimed at conserving and replenishing natural resources. The practical application of this technology and the role of naturally occurring populations of microorganisms in agroecosystems form the basis of this article.

Soil microbial biomass is composed of eukaryotic (fungi, yeasts, protozoa and algae), and prokaryotic (eubacteria, actinomycetes and archaea) organisms, whose populations vary from soil to soil. Many microorganisms possess urease enzymes which play a role in soil enrichment through the degradation or hydrolysis of organic nitrogen. Soil microbial biomass is dictated by a variety of soil and environmental parameters including soil texture and structure, pH, air/moisture content, and soil temperature. One of the prime factors that determine soil microbial status is the type and amount of organic material that enters the soil ecosystem. The vast majority of soil microorganisms are heterotrophic and require organic materials as both carbon and energy sources. Management practices in particular, the manipulation of the quality and quantity of organic inputs might, therefore, be predicted to modify soil microbial populations, the soil food web, and the biological processes involved in nutrient transformation. Microorganisms play a fundamental role in soil creation and stability through the binding of soil aggregates by hyphae and by the secretion of exudates. There are many examples of the ways in which soil microbial systems help ecosystem health and stability. In a study examining the influence of the microbial community of a Senegalese Sahel soil on the interactions between root knot nematodes and Pasteuria. penetrans, a parasite of plant-parasitic nematodes, the actinomycete was associated with a larger soil microbial population, including mycorrhizal and nematophagous fungi, which together stimulated the attachment of P. penetrans to the nematodes to reduce nematode infection.

Antagonistic and antibiotic microorganisms There are many reports that soil organisms might be antagonistic to plant pathogens, pests and weeds. For example, over 100 species of fungi trap and prey on nematodes, and many fungi are hyperparasites of other fungi. These activities not only influence the general nutrition, health, and vigour of higher plants (which also affects disease susceptibility), but they also determine the competitive behavior of root-infecting fungi and their microbial antagonists. Streptomycetes, filamentous bacteria that are effective and persistent soil saprophytes often associated with plant roots, are well-known producers of antibiotics and extracellular hydrolytic enzymes. They have the potential to contribute significantly to an integrated disease management system that includes alfalfa and other crops such as potato, maize and soybeans due to their ability to colonize plants and decrease damage from a broad range of pathogens. Mechanisms by which endophytes can act as biocontrol agents include production of antibiotic agents, siderophore production, nutrient competition, niche exclusion and induction of systemic acquired host resistance. Bacterial endophytes can thus play a role in pathogen suppression and complementary crop sequences can encourage beneficial allelopathy. Rhizobacteria, particularly the so-called plant growth promoting rhizobacteria (PGPR), can directly suppress plant pathogens by the production of antibiotics and hormones, and by competition with pathogens for resources. Some of these bacteria also promote root and shoot growth, nodule formation by Rhizobium (when they are known as nodule promoting rhizobacteria, NPR), and mycorrhiza establishment (as mycorrhiza helper bacteria, MHB).

Mycorrhizal microorganism Mycorrhizal fungi (microbial associations with mycorrhizal fungi and bacteria that live on and near the roots) offer several benefits to the host plant, including faster growth, improved nutrition, greater drought resistance, and protection from pathogens. Plant protection arises from the formation of a physical barrier to the invading pathogen, secretion of antagonistic chemicals, competition with the pathogen, increases in the nutrient uptake ability of plant roots, and by changing the amount and type of plant root exudates. Mycorrhizal fungi can be broadly categorized into two groups: vesicular arbuscular mycorrhizas (VAM or AM) and ectomycorrhizas (EM), which vary widely in structure and function. The former are microscopic obligate symbionts that can only be cultured with a host plant and constitute the largest proportion of fungal biomass in the soil. VAM produce structures known as arbuscules and vesicles within the root cell that are involved in the nutrient transfer process. This symbiosis confers benefits directly to the host plants through the acquisition of phosphate and other mineral nutrients from the soil by the fungus while the fungus receives a carbon source from the host. In addition, the symbiosis may also enhance the plants resistance to biotic and abiotic stresses. Ectomycorrhizas (EM), regularly produce large above-ground fruiting bodies (the familiar mushroom and toadstool) as well as a hyphal net around the plant root. Vesicles and arbuscules are absent and hyphal penetration of the root is limited. EM fungi are amenable to axenic culture. Potential inoculum can also be produced in the field from mycelium. It is generally agreed that in agriculture and forestry, the most significant function of both EM and VAM fungi is in optimizing phosphate uptake by the plant. Since P uptake by plants occurs at a faster rate than the movement of P in the soil, a P depletion zone can develop at the root surface. This is most significant in low P soils, as would be the case in many organic, low input or subsistence farming systems. The addition of P fertilizer to ostensibly low-P soils have the effect of increasing P availability to plants but it suppresses mycorrhiza. It follows that the long-term use of fertilizers, as is usually the case in modern agriculture, will have the effect of creating fertilizer dependency since the natural resources represented by soil organisms and the soil itself may be systematically depleted.

Mycorrhizal fungi are also believed to control root pathogens by one or more of:

- Improving nutrient acquisition by host plant;
- Excluding pathogens at root infection sites and within the rhizosphere;
- Inducing anatomical and structural changes in the root thereby creating physical barriers to pathogen entry;
- Producing substances antagonistic to root pathogens; and
- Activation of plant defense mechanisms.

The role of mycorrhizal fungi, as well as other microorganisms, in stimulating induced resistance to soil-borne plant pathogens in a range of crops has received much attention over the past 25 years and it is worth considering its significance to organic farming. Interactions between plants, mycorrhizal fungi and soil bacteria are not random but appear to occur in an ordered manner between compatible species. The presence in the soil of given mycorrhizal species can influence the persistence and activities of native bacteria and vice versa. Combinations of a range of beneficial bacteria and fungi have also been used not only to improve plant growth and health but also to achieve concomitant improvements in soil. In experiments to assess the effectiveness of combined mycorrhizal inoculation and compost application for afforestation with Olea europaea, significant improvements in soil structure (increased aggregate stability and decreased bulk density) were reported. The growth of O.europaea was significantly enhanced by both composted organic residues and mycorrhizal inoculation treatments. The increase in mycorrhizal O. europaea seedling growth was attributed to the positive influence of mycorrhizal fungi on soil aggregate stability. In addition, the positive interaction between composting and VAM inoculation that led to increased growth of tree seedlings was related to the ability of the mycorrhiza to enhance plant nutrient uptake from the compost. The possibilities for exploiting these types of synergism have been investigated in many cases. Heavy metal pollution of soils (along with severe compaction) is common on metalliferous mine spoils and on modern brown-field sites following industrial activities. Polluted soils such as these are slow to support new vegetation and so remain prone to further damage from erosion. One of the ways in which plants benefit from ectomycorrhizal infection in nature is through an increased tolerance to heavy metals in the soil. Mycorrhizal fungi confer protection to the plant against uptake of heavy metals by adsorption on hyphae and through various forms of fungal metabolism.

Organic Farming and Microorganisms

Soil management practices that change the soil carbon content also influence the size and structure of the soil biomass. Application of organic amendments and/or cover crops increases carbon availability to microorganisms. After more than 40 years, microbial biomass C was higher in organic than in conventionally managed dairy farm soils. Carbon released from crop residues contributes to increasing soil microbial activity and so increases the likelihood of competition effects in the soil. The placement of the residue in soil can lead to the displacement of a pathogen from its preferred niche, diminishing the pathogens' ability to survive. Therefore, soil microbial biomass changes as a consequence of switching to organic land management. Application of organic matter and all treatments that increase the total microbial activity in the soil might enhance the general suppression of pathogens by increasing competition for nutrients. Crop rotation may also yield microbial benefits beyond those normally associated with pathogen host range and saprophytic survival. Rotation is most successful in limiting the impact of biotrophic pathogens that require living host tissues, or those pathogens with low saprophytic survival capability.

Agricultural practices can have major short- or long-term impacts on mycorrhizal fungi, as well as on other soil microorganisms. Studied the effectiveness of VAM spores from organically and conventionally managed soils and found that white clover only benefited from mycorrhizal infection in a low-fertility (organically managed) soil. Furthermore, inoculate from organic soils were more effective in both achieving mycorrhizal infection and in allowing more efficient P uptake by the crop. Application of composts may enhance beneficial soil microorganisms. As the active microbial biomass increases, the capacity to utilize carbon, nutrients and energy in the soil is increased and thus these resources might be limited for soil-borne pathogens. It is clear that soil microorganisms play a fundamental role in soil health and sustainability. Microbial population density and diversity are affected by the level of organic matter indirectly by providing energy for soil microorganisms, improving soil structure and stability, soil moisture and plant nutrient availability and by preventing soil-borne disease incidence. Crop and soil management practices designed to support soil microorganisms i.e., minimum tillage, cover cropping, composting, eliminating pesticide and fertilizer use, and maintaining biodiversity, will inevitably lead to soil improvements and stability. The need for replenishing organic matter in the soil after harvesting is thus of great importance. In ecological farming systems, organic matter is maintained by mixed farming, rotations, recycling, compost and farm yard and green manures and bought-in organic sources. Including legumes and cover crops in the rotation should help balanced soil fertility, disease management, and avoid leaching nutrients and soil pollution. Crop and soil management practices designed to support soil microorganisms have been shown to lead to improvements in soil structure and stability.

Although the great diversity of soil microorganisms is now well appreciated, studies over the past twenty years have established that a number of other soil organisms, apart from mycorrhizal fungi, not only play an active role in suppressing pathogens but also mediate the activity of a range of beneficial organisms. Apart from farming in ways that support indigenous microbes, the use of proprietary cultures of beneficial microorganisms as soil and compost amendments, root dips, feed additives or sprays in order to supplement or reinforce the indigenous micro-flora are being considered.

Uses of Biofertilizer for Plants Promotion

The pattern of variation in plant upper growth, yield, and soil behavior under different treatments are illustrated by. To achieve successful and reproducible results following the introduction of beneficial rhizobacteria with organic matter and into soil, its survival in the heterogeneous soil environment is very important. Owing to the constrains associated with the inoculum's formulation, various introduction methods were attempted in the current study. The processing of microorganisms in various ways can have remarkable impact on microbial viability and field application. The success of the study may likely be due to the easy mixing of Pseudomonas fluorescens with organic matter without interfering with the environmental constrains. The observations on growth and yield parameters of Chili and Brinjal were recorded during the course of investigation and statistically analyzed and the significant of results were verified. Results with main effects and significant interaction have been described in details in succeeding graphically. To provide better understanding, some of the characters have been represented graphically. The result obtained was in accordance with previous studies where growth of cucumber, ladyfinger and pea was reported to be higher in treatments of biofertilizer combined with organic matter as a biofertilizer have conducted a series of laboratory experiments with bacterial strains on cucumber and reported an increase in plant height (38%), plant yield (36%), plant leaf quality (20%) and weight (13%) due to bacterial and organic matter effect.

Bottle gourd

Bottle gourd originated in Tropical Africa and domesticated in Asia, Africa and New World. India is the second largest producer of vegetables in the world after China. According to recommendation given by India Council of Medical Research (ICMR) an average man with vegetarian or non-vegetarian food habit should consume 300 g vegetables per day, which include 125 mg leafy vegetables, 100 g of root vegetable and 75 g of other vegetables. In the year 2002, the total vegetable production of the country was 97.5 million tons from 7.59 million hectares of land. In the country, vegetable crops are grown only in 2.8% of total cultivated land and share 10% of the world's vegetable production with the productivity of 13.6 t ha-1, which is quite low as compared to other advanced countries.

During summer the crop is sown in February-March and the vines are allowed to spreads on the ground. The rainy season crop is grown in unirrigated upland condition and sown in July- August; the vines are trained to climb on some artificial structure. The area and production of summer season crop is higher than the rainy season crop. In the temperate climate, the crop may be grown during summer-rainy season, if the temperature ranges from 150C to 250C available for 100-120 days. In tropical conditions, it can be grown throughout the year under irrigated conditions. Different parts of the bottle gourd have medicinal properties also. It contains 96.3% water, 2.9% carbohydrate, 0.2% proteins, 0.5% fat, 0.5% mineral matter, 11 mg of vitamin C. It is a highly cross-pollinated crop in which a large amount of variation is observed in many economically important traits. Nutrient Arrangement involves an efficient and judicious supply of all the major and minor components of plant nutrients. Chemical fertilizer in combination with farmyard manure, compost poultry manure, vermicompost, neem cake, biofertilizer crop residues or recyclable waste and other locally available nutrient sources for sustaining soil fertility, health and productivity of the soil. The integrated supply and use of plant nutrients from organic fertilizer have been proved to produce higher crop yield than when each is applied alone in bottle gourd.

Besides soil climatic factors and cultivars, it is not sufficient for analysis of disease resistance and best quality fruit and seed for further utilization. The majority of farmers are still using fertilizer Urea, DAP, MoP and FYM for giving recommended doses of fertilizer and organic manure but not in a particular proportion so there is a need to search a required dose of organic fertilizer. The successful cultivation of Bottle Gourd under agro forestry system required information assessment of the system. Keeping in view of the importance of the crops, an investigation was made to study the effect of organic farmers on the growth and yield of Bottle Gourd.

Brinjal

The common large-fruited forms are believed to have originated in Indo-Burma region. Fruits are moderate sources of vitamins and medicinal properties including de-cholestrolizing action. China is the largest producer of Brinjal and contributes about 68.7% of the world's Brinjal production while India occupies second position in production with a share of 23.3%. However, the productivity of Brinjal is quite low in India. In India, Brinjal occupies fourth position in area and sixth in production among the vegetable crops. Shoot-root borer, bacterial wilt, fusarian wilt, little leaf are the major threats to the brinjal. The use of organic solution (Biofertilizer) not only spoils the ground water, soil but also have deleterious effects by the emission of harmful gases. Forum for Nuclear Co-operation in Asia Bio-fertilizer Project, National Project on Organic Farming and All India Network Project on Bio-fertilizers aims to encourage use of bio fertilizers. Bio-fertilizers improve the quantitative and qualitative features of many plants. Biofertilizers used in conjunction with chemical fertilizers improve crop productivity and nutrient use efficiency. Positive effect of Pseudomonas fluorescens on growth and yield of Brinjal also been reported by many workers. There is a positive influence on the growth and yield attributes of Brinjal.

It is becoming difficult to meet the nutrient need of farming through chemical fertilizer alone and due to its higher costs; the concept of integrated plant nutrient supply system (IPNS) is gaining ground. Therefore, the investigation was planned and conducted to study the influence of liquid biofertilizers alone, dual and in different combinations with chemical fertilizers on the vegetative and reproductive growth, yield and quality attributes of Solanum melongena L.

Chilli

Chilli (Capsicum annuum L.) is the universal spice and widely cultivated throughout temperate, tropical and subtropical countries. It is an indispensable spice due to its pungency, spice, taste, appealing odor and flavor. The dried fruit are extensively used as spices. The ground powder and oleoresin are utilized in pharmaceutical preparations. Nutrient management is one of the most important factors to improve the productivity of chili. The continuous sole and erratic use of chemical fertilizers in imbalance form leads to decline in soil fertility as well as nutrient uptake efficiency of plants, resulting in either yield stagnation or decrease consequently. In last four decades many microorganisms have been used in the form of bio fertilizers. Bio fertilizers are eco-friendly, low-cost input and not only improve the crop growth and yield but also improve fruit quality and fertilizer use efficiency. Bio fertilizers are the preparations containing living or latent cells of efficient strains of microorganisms capable of fixing atmospheric nitrogen, solubilizing phosphorous and potassium, mobilizing nutrients and absorptions of water, decomposing cellulolytic and alginolytic water materials. These microorganisms also produce the growth promoting substances which greatly influences the seed germination, root growth and proliferation.

The Pseudomonas fluorescens from its role in nitrogen fixation from the atmosphere is also involved in production of bioactive substances. Application of phosphate solubilizing microorganisms has given promising response in improving yield of various crops. Mycorrhizal fungi are the distinct symbiotic association between plant roots and soil fungi. Potassium solubilizing bio fertilizers are able to solubilize potassium minerals through production and excretion of organic acids. To maintain long term soil health and productivity there is a need for integrated nutrient management through manures and bio fertilizers. Therefore, the current trend is to explore the possibility of supplementing chemical fertilizers with the bio fertilizers. Keeping in this view the present investigation was undertaken to find out the effect of bio fertilizers with graded levels of inorganic fertilizers on growth and yield of chili.

Cucumber

Cucumber (Cucumis sativus L.) is an important vegetable and one of the most popular members of the Cucurbitaceae family. It is thought to be one of the oldest vegetables cultivated by man with historical records dating back 5,000 years. The crop is the fourth most important vegetable after tomato, cabbage and onion in Asia, the second most important vegetable crop after tomato in Western Europe. In tropical Africa, its place has not been ranked because of limited use. Fertile soils are used for the cultivation of cucumber; infertile soils result in bitter and misshapen fruits which are often rejected by consumers. Bush fallowing has been an efficient, balanced and sustainable agricultural system for soil productivity and fertility restoration in the tropics, but as a result of increase in the population, the fallowing periods have decreased from ten years to three years and this has had an adverse effect on the fertility restoration leading to poor yields of crops. Therefore, the use of external inputs in the form of farmyard manures and fertilizer has become imperative. Farmyard manure has been used as a soil conditioner since ancient times and its benefit have not been fully harnessed due to large quantities required in order to satisfy the nutritional needs of crops. The need for renewable forms of energy and reduced cost of fertilizing crops, have revived the use of organic manures worldwide. Improvement in environmental conditions and public health are important reasons for advocating increased use of organic materials. However, because it is bulky, the cost of transportation and handling constitute a constraint to its use by peasant farmers.

Farmyard manure release nutrients slowly and steadily and activates soil microbial biomass. Organic manures can sustain cropping systems through better nutrient recycling and improvement of soil physical attributes. The use of inorganic fertilizer has not been helpful under intensive agriculture because of its high cost and it is often associated with reduced crop yields, soil degradation, nutrient imbalance and acidity. The complementary use of organic fertilizers has been recommended for sustenance of long-term cropping in the tropics. Nutrients from mineral fertilizers enhance the establishment of crops while those from mineralization of organic solution promoted yield when both fertilizers were combined. The most satisfactory method of increasing maize yield was by judicious combination of organic wastes and inorganic fertilizers. It has been observed that addition of manure increases the soil water holding capacity and this means that nutrients would be made more available to crops where manures have been added to the soil observed that nutrient use efficiency might be increased through the combination of organic solution. This study was therefore conducted to investigate the effects of varying rates of farmyard manure and organic solution (Biofertilizer) on the growth and yield of cucumber.

Lady finger

The plant to plant and row to row spacing were maintained at 30x60 cm respectively. The experiment was carried under Randomized Block Design (RBD) with three replications. A single standard variety (Kasha pragati). On the basis of overall performance under the present investigation, it may be concluded that the application of Pseudomonas fluorescens + Organic solution in Kashi Pragati increased the growth, yield and nutritional quality of okra. It was, therefore, concluded that the use of organic manure in combination with inorganic fertilizers in the production of vegetables like okra should be encouraged as it is beneficial for the physical growth of okra plant. Okra or bhindi belongs to the family Malvaceae Okra [Abelmoschus esculentus L. (Moench)], is an economically important vegetable crop grown in tropical and sub-tropical parts of the world. It is generally an annual plant. It is also known as lady finger or bhindi, originated in tropical Africa. Because of its richness in nutrition, taste, medicinal and industrial value okra is one of the most popular vegetables in all section of people. Okra is cultivated for its fibrous fruits or pods containing round, white seeds. The fruits are harvested when immature and eaten as a vegetable. Okra is mainly propagated by seeds and has duration of 90-100 days. Okra can be grown in both the seasons i.e., Rabi and Kharif. Okra requires a long, warm and humid growing period. Okra is a good source of vitamins, minerals, calories, and amino acid found in seeds and compares favorably with those in poultry, eggs and soybean.

All parts of okra (Ladies' fingers) like fresh leaves, buds, flowers, pods, stems and seeds can be used for different purpose and hence it is a multipurpose crop in term of its use. The mucilage found in okra may be used for plasma replacement or blood expender. Moreover, long term sustainability of productivity could be achieved only through the interaction of inorganic and organic sources of nutrients. Nitrogen is a single most important nutrient which contributes to the proper growth of plant and yield. Organic manures improve the quality of green pods. Therefore, the applications of plant nutrients through organic sources like compost, farm yard manure and biofertilizers remains the alternative choice of the growers for maintaining its sustainable production. Indiscriminate use of inorganic fertilizers has resulted in decreased nutrient uptake, poor quality of vegetables and deterioration of soil health. Okra produces fruit for a long time and needs balanced and sufficient supply of nutrients for higher yield and better quality) the objective of this experimental study is to evaluate the performance of integrated nutrient management on certain yielding, fruiting and flowering qualities parameters by using okra.

Potato

Potato (Solanum tuberosum L.) is the world's fourth most important crop, after maize, wheat and rice, organic fertilizer is one strategy to increase tuber yields in potato. Due to consumer concern for excessive use of chemical fertilizers to avoid deleterious effects on the environment, there is increasing interest in the use of organic fertilizers. Organic fertilizers provide a slow release of nutrients as micro-organisms in soil. However, use of organic fertilizers is limited by the threshold of nitrogen level, which is allowed, especially for exportation of organic products. Biofertilizers from microorganisms can replace chemical fertilizers to increase crop production. In principle, biofertilizers are less expensive and are more environmentally friendly than chemical fertilizers. The benefits from using biofertilizers, beside their role in saving chemical fertilizers are well known. The present work aimed to study the effect of using 100% organic matters and bacteria (pseudomonas fluorescens) on potato growth and yield as compared to the conventional mineral fertilization.

Ridge gourd

Ridge gourd (Luffa acutangula) is one of the most important warm season vegetables which are commercially propagated by seeds. It contains high content of water and nutrients, protein, fat, carbohydrates, minerals and vitamins. The treatment of seeds with biofertilizers or the soil application of biofertilizers does not pollute the soil and also does not show any negative effect to environment and human health. One of the major essential elements for growth of plants is nitrogen and it is provided in the form of synthetic chemical fertilizer (urea). Such chemical fertilizers pose a health hazard and microbial population problem in soil besides beings quite expensive and making the cost of production high. In such a situation the biofertilizers play a major role.

The biofertilizers, are alternate low-cost plant nutrient resources which play a vital role in maintaining long term soil fertility substances. The Biological nitrogen fixing microorganisms significantly contribute for nitrogen addition to soil; phosphate solubilizers help in solubilize bound form of phosphorous. These beneficial microorganisms are known to secrete plant growth promoting substances. Biofertilizers are the formulation of living microorganisms, which are able to fix atmospheric nitrogen in the available form for plants either by living freely in the soil or being associated symbiotically with plants. Biocontrol fungi of the genus Trichoderma fungi can utilize a variety of nutrient sources are able to effectively degrade some of them biofertilizers function as key player in sustainable agriculture by improving soil fertility, plant tolerance and crop productivity current soil management strategies are mainly dependent on inorganic chemical-based fertilizers, which caused a serious threat to human health and environment. The exploitation of beneficial microbes as biofertilizers has become paramount importance in agriculture sector for their potential role in food safety and sustainable crop production.

Tomato

Tomato (Lycopersicon esculentum Mill) is one of the most popular and widely grown vegetable in India. It covers 7.9% of total area grown under vegetable crops and 9% of total vegetable production. Various factors are responsible for its low yield; inadequate use of chemical fertilizers is one of them. The crop particularly hybrids have higher requirements of nutrients. Low use of biofertilizers and imbalance responsible for the low yield. Moreover, complete dependence on chemical fertilizers is not sufficient to sustain the higher yields. Also, due to high cost of inorganic fertilizers, sometimes the margin of profit remains poor. The use of biofertilizers may be the right answer to above mention problems. Biofertilizers, which are ecofriendly and more economic, can play an important role in reducing the dependence on chemical fertilizers. Biofertilizers fix appreciable amount of atmospheric nitrogen in soil, enhance plant growth by production of organic acid and growth hormones and make available the complex phosphorus to the plant, which may cause an appreciable reduction in consumption of inorganic fertilizers. Keeping these in view, response of inoculation of biofertilizers on growth and yield of tomato were carried out under field condition.

4

The Environmental impacts of Organic Farming

A high proportion of the Earth's land surface has been transformed by direct human action, with significant consequences for biodiversity, nutrient cycling, soil structure and biology, and climate. Agriculturally used land accounts for 38% of the world's total land area, of which 69% is permanent pasture, making it the leading type of agricultural land use; 28% is arable land and 3% is under permanent crops (FAO 2004). The location and physical production conditions are important determinants of the types and intensities of agricultural land use; in particular, the management of relatively flat, fertile land has often been progressively intensified, with mechanization leading to increased field sizes, removal of boundary vegetation and increased application of agrochemicals. In contrast, traditional farming systems on marginal land, where possibilities for mechanization are limited because of steep or inaccessible terrain, are in many areas under considerable economic pressure or have been abandoned already. The term Organic Farming used in this chapter is based on the Codex definition of the FAO. Organic Farming, which relies on natural methods of building soil fertility and controlling pests and diseases, can be divided into two categories: certified organic farming and de facto organic farming. Whereas certified organic production is often oriented towards the market for food labeled as organic, the latter may represent the major part of organically managed land. Such de facto organic farming can often be found in resource-poor and/or agriculturally marginal regions where local populations have a limited engagement with the cash economy. This includes many traditional farming systems found in developing countries, which have evolved through centuries to create agricultural systems adapted to local environmental and cultural conditions. This chapter covers the environmental impacts of organic farming in this wider sense and also includes agro forestry, a management system that integrates trees in the agricultural landscape and that is systematically applied in Organic Farming. Agro forestry is practiced in all agro climatic zones but is most prevalent in the tropical belt.

Organic Farming is generally perceived as a form of agriculture that is more favorable for the environment than conventional agriculture. In this chapter it is asked where, precisely, the differences are in environmental terms and whether the available empirical evidence allows for a generalization. Geographical differences and interrelationships are emphasized wherever relevant. Primary sources for this chapter are a review of research results, mainly for developed countries in temperate and subtropical zones. There is a lack of scientific research in developing countries in the tropics and subtropics, but from practical experiences and case studies, it may be assumed that results are transferable.

A wide range of indicators are being used in the description of the environmental impacts of agriculture. Particularly, the organization for Economic Cooperation and Development (OECD) pushes the development of a common set of indicators at an international level. This framework is internationally accepted and allows cross-country comparisons. Whenever appropriate, simplifications and modifications have been made. The main indicators used in the analysis presented in this chapter are given in table 4.1. The discussion of environmental impacts are structured by six main sectors. There are significant interrelationships between these sectors.

Table 4.1: Overview of the main fields of environmental impact examined and the specific aspects covered (after OECD 1997)

Area	Aspect/indicator
Biodiversity	Genetic diversity Floral diversity Faunal diversity Habitat diversity
Landscape	Landscape structures and aesthetic value
Soil	Soil organic matter (SOM) and acidity (pH value) Biological activity Soil structure Erosion Desertification
Ground and surface water	Nutrient use and balance Nitrate leaching Phosphorus Pesticides Pathogens
Climate and air	Carbon dioxide (CO_2) Nitrous oxide (N_2O) Methane (CH_4) Ammonia (NH_3)
Energy	Intensity of energy use Efficiency of energy use

Biodiversity

Since its beginnings, agriculture has been a source of positive and negative effects on the eco-system in terms of wildlife conservation and landscape. The level of intensity that modern agriculture has reached in wide areas of Europe has resulted in a decrease in biological diversity of domestic and wildlife species. The destruction of biotopes, as well as the simplifications of crop rotations, and the increasing input of synthetic fertilizers and pesticides have been responsible for changes in the habitat of many species. Maintaining and enhancing biodiversity is considered central to developing a sustainable organic system. Organic farming depends upon stabilizing agro ecosystems, maintaining ecological balances, developing biological processes to their optimum and linking agricultural activities with the conservation of biodiversity. Increased biodiversity improves and buffers ecological services such as pollination, pest control, maintenance of soil fertility, thus strengthening farming systems and practices. Building on that, some organic certification organizations have incorporated biodiversity requirements into their standards. The Swiss organic standards, for example, require farmers to use 7% of their land as semi-natural habitats.

Organic farming systems seem to be an appropriate tool for planners to balance conservation and production, but the philosophy behind the more biodiversity the better requires a deeper discussion. Many researchers have claimed that processes and functional groups of organisms are more important for ecosystem function than just maximizing the diversity. In a review of 76 studies that explicitly compared the effects on biodiversity of organic farming relative to conventional agriculture, highlighted three broad management practices that are largely intrinsic to organic farming (but not exclusive), and particularly favor farmland wildlife: a ban or a reduced use of chemical pesticides and inorganic fertilizers, sympathetic management of non-cropped habitats, and preservation of mixed farming. While comparing the impacts of organic and conventional farming systems on biodiversity, identified the following problems:

- Variation in the definition of organic farming standards between countries;
- Disparity between studies in their control for extraneous variation;
- Variation in the period the studies were carried out;
- Variation in the spatial scale; and
- Use of different measures of biodiversity.

Furthermore, there are several factors that may result in underestimating the benefits of organic farming on biodiversity, such as a possible time lag in the response of wildlife communities after switching from conventional to organic farming, or the difficulty in detecting significant effects at the field scale for mobile taxa such as birds and butterflies. Nevertheless, most studies reviewed clearly demonstrate that species abundance and/or richness across a wide range of taxa tend to be higher on organic farms than on locally representative conventional farms. This particularly applies to species which have experienced declines in range and/or abundance as a consequence of past agricultural intensification. The effects on biological diversity are reviewed in more detail for the following three aspects: genetic diversity, species diversity (floral and faunal) and habitat diversity.

Genetic Diversity

Today, the adoption of high yielding, uniform breeds and varieties has led to a considerable reduction in the number of species and in the number of varieties/breeds within species used in agriculture. There are many schemes and projects worldwide working to conserve seed banks and indigenous varieties, many of which are linked to Organic Farming projects. A typical example is the Sustainable Agriculture and Rural Development Project (SARDI) in Kenya where a community indigenous seed conservation program is being implemented. Indigenous seeds have been shown to perform better in drought conditions. Indians in the High Jungle of Peru who organically cultivated cotton and thereby conserved ecotypes adapted to the moist tropical conditions. Vreeland emphasizes that native cotton can also provide valuable genetic materials for improving commercial cotton varieties because of the high degree of natural resistance to insects, disease and drought stress. Even though there is evidence that organic farming systems are more likely to use rare, native or traditional breeds, there are only few studies that investigate the role of organic live-stock in maintaining the genetic diversity of domesticated stock, since 1995, Organic Farming has indirectly established a rescue process for species, varieties and breeds threatened by underuse or extinction. As demonstrated in various case studies, Organic Farming, almost without the help of governmental institutions, is providing an important contribution to the in-situ conservation, restoration and maintenance of agricultural biodiversity. A possible future threat to genetic diversity and biodiversity in general, could be the side effects of the release of genetically engineered or modified organisms (GMOs) into the environment. In Organic Farming genetic engineering is banned.

Floral Diversity

In general, the diversity of floral species is closely connected to local site conditions. Nevertheless, while investigating the flora of arable and mixed farming, recorded higher weed abundance and species richness in fields under organic management in almost all studies. In several studies, organically managed fields held considerably rarer and/or declining species. However, in regions with low potential for biodiversity, the positive impact of organic farming on wild herb or grassland diversity is less distinct. Factors that contribute to greater floral diversity include the proportion of grassland relative to arable cropping, the variation in sowing dates for cereal crops and the inclusion of both autumn-sown and spring-sown cereals. A richer floral diversity has positive impacts on faunal diversity, because it offers overwintering sites, refuges and areas with network links to other habitats. Weeds are considered competitive to the crop in conventional farming and are eliminated by herbicides, whereas in organic systems, some of the accompanying plants are desired to a certain degree and are considered useful as they provide ecological services. In the context of pollinators, which greatly benefit from a diversity of flowers, flowering weeds are more diverse and abundant in organic arable fields and grassland compared to conventional fields. In organic grassland, the average number of species was higher than in conventional, as a result of lower stocking rates and lower fertilization levels in organic farms. Additionally, the mowing date is often delayed, which means that grass species can reach the flowering stage and thus achieve greater reproduction, leading to plant communities rich in species and structure.

They found the farmland area dedicated to natural elements ranged from 55% to 20% in the mixed organic farms and from 11% to 0.3% in the conventional neighbors. The conservation value of hedge bottom vegetation on organic and conventional farms in Denmark. Studied were established in the same way except that the organic hedgerows were managed without pesticides. Significantly more plant species were found in the organic hedgerows and their species composition appeared similar to semi-natural plant communities. The biological relationship between floral diversity and both pest and beneficial insects has been studied extensively. The importance of flowers to attract beneficial insects for adult feeding as well as grass banks as overwintering sites for predatory beetles is now fully recognized. Uncultivated land, hedgerows inside crops, in allied crops and in neighboring areas can serve as refugia for beneficial insects such as parasitoids of aphids. An organically farmed experimental system, and indicated that short rotation coppice hedges and clover/grass leys in the rotation may increase diversity and activity of parasitoids attacking cereal aphids. Besides the positive effects of biological pest control, from a farmer's point of view, negative interactions such as pests overwintering in hedges and weeds invading from field margins are also important. Many farmers considering converting to organic farming fear that such problems will be very difficult to manage. Although some mechanisms of natural enemy enhancement, such as increasing diversity within cropping systems, have been explained in specific systems, it is difficult to show universal trends due to the great diversity of systems and competing processes that vary under different conditions.

Concerning domesticated species, organic farming aims for a greater diversity of crops in their rotation. Wide crop rotations are essential as a means of disease and pest prevention. They also contribute to maintaining soil fertility, particularly if N-fixing legumes are part of the rotation. While organic farming standards recommend cultivating site adapted crop varieties, organic farmers (just like their conventional colleagues) often choose modern high yielding varieties. Nevertheless, the preservation of old land varieties and breeds is an important initiative within the organic farming movement.

Faunal Diversity

The effects of organic farming on faunal biodiversity have been studied, particularly for soil fauna and birds. In most cases, organic farming displays a greater faunal biodiversity than conventional farming. Key factors are: greater fauna friendly crop protection management, the organic fertilization regimes, the more diversified crop rotation, and the more structured landscapes with semi-natural habitats and field margins. Landscape structures are essential for the survival of many invertebrates, especially due to favorable food and overwintering conditions. They also function as habitat crosslink between meadows, fallows and field margins. An analysis of some specific groups of fauna illustrates the interrelationships.

Earthworms

Earthworm populations can indicate the structural, microclimatic, nutritive and toxic situations in soils. They are highly suitable bioindicators of soil fertility, and they are known for their sensitivity to synthetic pesticides and to many agricultural practices. Earthworms help to improve soil structure and provide a high concentration of nutrients in a form accessible to plants. Many investigations in Europe and North America show that organically managed soils tend to exhibit a greater abundance and species number of worms compared to conventionally managed plots or farms.

Arthropods

Pesticides can affect beneficial arthropods either directly via contamination, or through alteration of the microhabitat and a reduction of their prey. In contrast to conventional farming, synthetic pesticides are banned in organic farming and the use of a few approved natural pesticides is very restricted. Most natural pesticides are not very selective and can therefore also affect beneficial arthropods. But the toxicity of natural pesticides is usually lower and their degradation significantly faster in comparison to synthetic pesticides. Additionally, the higher fertilization levels in conventional fields lead to a greater density of crops, which can alter the microclimate and can also reduce the occurrence of species dependent on a warm microclimate and on light demonstrated that tomato plants fertilized with high nitrogen (N) levels showed a higher intrinsic rate of increase of greenhouse whitefly populations. studied the effects of organic and synthetic fertilizers on pests (aphids) and predatory arthropods associated with tomatoes. The application of both organic and synthetic fertilizers could increase pest populations. However, they found lower populations of aphids on organically fertilized tomatoes after one year, indicating that organic fertilizers may have the potential to reduce pest attacks in the long term. In Europe, researchers have found greater diversity and abundance of soil and surface - living arthropods such as spiders, beetles, parasitic flies and wasps as well as non-pest butterflies, and many other invertebrate species in organic farming systems compared to conventional farming systems. However, some reports have not found these differences to be as clear, and in the case of aphids, there are generally fewer present in organic fields examined the abundance and ecological diversity of selected groups of beneficial arthropods on 16 organic and 17 conventional carrot fields in New Zealand. The organic fields had significantly higher numbers of Hymenoptera and Neuroptera as well as a significantly more diverse predatory and parasitic community than the conventional fields.

Birds

The decline in farmland bird populations is well documented. Birds are well-suited indicator organisms that show the environmental status of nature and landscape infrastructure including agricultural land. Several studies show that bird densities are higher in organic farms. In coffee production in Latin America, another factor appears more important research carried out by the Smithsonian Migratory Bird Centre in Colombia and Mexico showed more than 90% fewer bird species in sun grown coffee plantations as opposed to shade-grown coffee. Shade-grown practices are recommended in organic standards as they fulfill requirements to enhance soil fertility, pest and disease control and expand crop production options. While many interactions between organic farming and biodiversity are positive, some actions can have a negative impact, particularly on ground-nesting birds. These include mechanical weeding, mulching operations, certain mowing techniques and timing of operations.

Mammals

There are only two strictly comparative studies of organic and conventional farming and their influence on mammalian biodiversity. Activity levels of small mammals (wood mouse, bank vole, common shrew) were greater in organic than conventional fields. Increased food abundance through hedgerows and field-edge habitats was cited as the most significant factor that benefitted these and a range of other mammalian species occurring on farmland. Many small mammal species are likely to have been affected by reduced insect and weed-seed food resources resulting from intensification investigated bat activity and species richness in a paired (organic/conventional) farm. Total bat activity was significantly greater on organic farms than on conventional farms, while no significant difference in species richness was found between the farm types. Responsible factors for the greater activity levels appeared to be better water quality and greater prey availability.

Comparatively few data are available on the effects of organic farming on other wild animal groups, particularly in regard to pollinators. This could be an important research theme. Apart from wild bees and other insects, bats are the principal pollinators of fruit trees and major staple food crops, including potato, cassava, yams, sweet potato, taro, beans, coffee and coconut. Declines in populations of pollinators now threaten both the yields of major food crops and the survival of wild plant species. Many vertebrate pollinator species are threatened in Australia, Colombia, Ecuador, Indonesia, Madagascar, Mexico, Papua New Guinea, Peru and the USA. The main reasons include: loss of nesting and roosting sites, habitat fragmentation by excessive exposure of nectar plants to herbicides and pollinators to pesticides, overhunting, disruption of nectar corridors required by migratory pollinators, and competition by invasive species.

Habitat Diversity

Semi-natural areas are extremely valuable habitats in the agricultural landscape with respect to the conservation of threatened species. They are also of great functional importance for nutrient cycling (e.g., meadows retaining nutrients and thereby preventing leaching) and processes of succession (e.g., colonization). In Denmark, there are four major pressures which affect semi-natural habitats such as permanent grasslands and meadows: changes in hydrology, fragmentation, eutrophication, lack of management and consequently, succession into shrub and forest argue that organic farming may affect these biotopes in the same manner as conventional farming except for the absence of chemical fertilizers and pesticides. While pesticides are rarely used on semi-natural areas in general, organic nutrients may be applied by organic farmers at (almost) the same levels as in conventional farming systems. Attention should therefore be given to developing organic farming practices that support the characteristic balance of functional groups of organisms, and that ensure a proper functioning of the relevant processes and functions. An example is the maintaining of unfertilized buffer strips along uncultivated biotopes.

Overall, there is little information available to compare habitat diversity in organic and conventional farming systems. Semi-natural habitats are intrinsic in organic regimes where their management is central to the philosophy. Organic farming does indeed tend to have a positive impact on habitat diversity, presumably for the reasons that have already been given. But the correlation is not very strong, since habitat diversity depends highly on given/historic landscape structures and site-specific aspects too. Diversity and density of habitats in the agricultural landscape may influence the production system, but the overall effects are extremely difficult to determine. There is a great need to intensify research of these interactions and to develop reliable indicators of supposed functional integrity and both the positive and negative interactions between cultivated and uncultivated areas in organic farming systems.

Landscape

Landscapes are territorial or spatial units produced through the interaction between human societies and cultures with the natural environment. They are viewed as integrating various functions demanded by society the multifunctional landscape. In this sense, landscapes can be classified according to their intrinsic beauty, historical features, embodiment of cultural values, past and present impacts of land use, farm practices, composition of farming systems, distribution of habitats and human made features like stone walls or historic buildings. Landscape goals and the relevant land use and management practices need to be addressed at the farm as well as at the landscape level. Suggested that a minimum of 3% to 5% of each farm should be designated to on-farm nature conservation, although the spatial separation of the conservation areas from the production zones should be minimized. State that minimum standards on the extent of uncultivated areas should rather be implemented at the landscape level, where due consideration of landform and habitat structure could be taken. More fields of smaller size and a significantly larger share of inner and outer hedgerows on organic farms are creating a more diverse mosaic within the farm found that the diversity of landscapes and production systems was greater in organic farms than on conventional farms, regarding land use types, crops, livestock, plantings (hedges, solitary shrubs, trees), flora and sensorial information. Farms exert a joint but uncoordinated impact on landscape. The development of high-input systems has reduced the area of extensively cultivated fields, grassland, uncultivated areas and small biotopes. Therefore, the landscape level management of habitats becomes crucial for the purpose of leaving room for natural processes. Development of organic farming in this respect is a potential instrument for such management.

Organic farming generally provides a good potential for landscape diversity, including criteria related to perception and sensory qualities. States that the basic principles of organic farming provide a perspective for further development of high-quality landscapes which include the possibility of a cautious utilization of sensitive areas, as well as a prequalifying of the identity of rural sites. How precisely organic farms (can) contribute to landscape quality is a main area for further research.

Soil

Soil care is a main principle in organic farming. It is, therefore, not surprising that the impacts of organic farming on soil properties have been researched comprehensively. Special focus in this research is on organic matter content, biological activity, nutrient cycles and soil erosion. Even though soil performance is very site specific, results of studies in different countries (e.g., Europe) show that organic farming tends to conserve soil fertility and system stability better than conventional farming systems. As organic farmers cannot use synthetic substances (e.g., fertilizers, pesticides, pharmaceuticals), they pay particular attention to operating a sound rotational system to nourish the soil in order to maintain organic matter content and to keep it in good condition. Organic management focuses on nutrient cycling with the aim of maximizing agro ecosystem stability and homeostasis. To restore the natural ecological balance is seen as essential by organic farmers because ecosystem functions are considered to be the main productive input.

Soil Organic matter

The environmental relevance of organic matter content is based on its capacity to improve nutrient availability as well as biological activity and to reduce the vulnerability to physical damage. Soil organic matter strongly influences many soil properties including bulk density, water holding capacity, infiltration rate, hydraulic conductivity and aggregate stability. Research on organic matter concentrates on soil organic carbon (C) content as a key parameter. The level of organic matter content correlates with site-specific conditions such as soil type, texture and precipitation as well as farming system. Soil organic matter (SOM) and humus are important components in the organic farming philosophy. Fertilization is based on organic substances such as farmyard manure from animal husbandry, compost, green manure, plant residues and commercial organic nitrogen fertilizers. Consequently, there is an extensive supply of organic matter passing through aerobic decomposition processes. Mineralization and decomposition processes are influenced by humidity, temperature and oxygen. This means that under tropical conditions these processes run faster and all year long, whereas under temperate conditions they are slower and come to a halt during the colder months. Several long-term trials that compare organic farming to conventional farming have been performed in various European countries. The research shows that soil organic carbon content is higher in organic systems than in conventional farming. As for pastures, the differences are less pronounced. However, some authors could not report an increase of SOM in organically managed soils. Soils in England under mixed organic arable rotations-maintained concentrations of SOM at similar levels to those under typical conventional systems.

The most important farm practices for organic matter supply vary in different regions as European organic farm characteristics differ considerably between climatic zones. Organic farming in the Northern countries is characterized by a high percentage of leys in crop rotations because animal husbandry is the dominant farm type. SOM content and composition on organic farms in the Mediterranean countries is based more on plant residues and green manure. Overgrazing and the need for fuel (both wood and dung) in the countries of Africa, Asia, Latin America and the Caribbean can both exert extreme pressures on local bio productivity, leading to nutrient losses and reducing the amount of available compostable and recyclable organic materials. Organic and agro ecological systems do not provide panaceas for areas with depleted and declining nutrient status. Organic and agro ecological systems can significantly help address problems of declining soil fertility by building up local productive capacity (both ecological and social), rather than relying upon external inputs.

Acidity and pH levels

The pH of the soil is an important parameter since it can affect the plants' ability to take up nutrients and the microbial activity in the soil that influences the processes required for plant nutrition. Changes in soil pH occur by the displacement of cations or by additions of sources of acidity such as hydrogen and aluminum ions. Chemical fertilizers are highly reactive and can cause extreme pH fluctuations in localized areas such as those near the fertilizer band. In contrast, organic manure can increase the buffering capacity of soils, preventing swings in pH, because of additional organic matter. Comparative studies of conventional and organic farming systems revealed that organic systems sometimes have higher pH levels in mildly acidic soils than their conventional counter parts. During a long-term organic farming study in the Great Plains in found that the pH levels in the alkaline soil decreased significantly over time with organic production. Results from a long-term Swiss trial comparing biodynamic, organic and conventional farming systems, the DOK trial, demonstrate that the utilization of composted manure, common in organic systems, has a positive effect on the content of organic matter and helps to avoid soil acidification.

Biological activity

High biological activity within the soil promotes metabolism between soil and plants and is an essential part of sustainable plant production and fertilizer management. The role of soil organisms is central to soil processes and fertility since they render available the elements in plant residues and organic debris entering the soil.

Earthworms and Mesofauna

Earthworms have many positive direct and indirect effects on soil quality, both in terms of their effects on soil physical properties and nutrient cycling. Furthermore, they are important in soil organic matter turnover. As a key species of soil mesofauna, they are an appropriate indicator of soil biological activities as a result of their sensitivity to any kind of soil disturbance. In a synthesis of relevant scientific results, compared organic and conventional farming systems and concluded that in organically farmed soils, a significantly higher biomass and abundance of earthworms occurred as well as a considerably higher diversity of earthworm species. During a long-term field trial and by others. A possible reason for the abundance of earthworms in organic farming could be that organic production depends more on a high, sustained supply of organic substance from plant residues and manure than conventional farming, which can rely at least partly on the mineral supply of nutrients. Organic rotations and particularly the inclusion of grass leys, preferably of several years (>2 years), into farming systems tend to favor earth- worms because of the beneficial effects of organic matter additions and leys. However, organic farming systems rely more on mechanical weed control and in certain crops, on intensive soil tillage as the use of synthetic herbicides is prohibited. This can have negative effects on other key species of soil mesofauna such as Collembolan insects (springtails).

Soil Microorganisms

The soil microbial biomass serves as a labile source and an immediate sink of carbon, nitrogen, phosphorus (P) and sulfur (S) in soils and performs critical functions such as nutrient transformation and pesticide degradation. Additionally, microorganisms form symbiotic associations with roots, control plant pathogens and participate in soil formation. In order to characterize soil microbial activity, parameters like total microbial biomass, diverse enzymatic parameters, carbon turnover and mycorrhization (mycorrhizal soil fungi build up symbioses between fungus and plant) are used. Generally, pesticides affect the population of microorganisms and fungicides tend to inhibit or kill soil fungi, including mycorrhizae which are particularly important in organic systems. At the same time, however, the evidence for increased microbiological activity under organic farming is mixed. In comparison to conventional plots, several scientists found higher microbial biomass, higher microbial activity and diversity as well as a higher efficiency in organic carbon turnover in organic plots reviewed 14 studies that investigated microbial communities under organic and conventional systems and found only limited differences. However, they detected a general trend towards elevated bacterial and fungal abundance/activity under organic systems.

The parameters of biological activity were influenced to different degrees by types of organic fertilizers dependent upon type and quality of the applied manure and agronomic techniques (crop rotation, soil tillage). Found in the Sustainable Agriculture Farming Systems (SAFS) Project (USA) that after 4 years, microbial biomass levels were consistently higher in organic and low input systems than in conventional systems, while plant parasitic nematode numbers were also consistently lower. Concerning soil fungi, found a larger number and abundance of saprophytic soil fungi with a higher potential of decomposition of organic material. Furthermore, the degree of mycorrhizal root colonization was found to be significantly greater in organic plots than in conventional plots. They found no consistent differences between organic and conventional farming. A possible explanation for the differing results could be that the level of biological activity changes very slowly in response to altering fertilization levels and cultivation techniques. This could explain why no differences in microbial activity between organic and conventional plots were observed in several on farm investigations. Several authors state that any experiment trying to assess these changes requires 8-10 years of post-conversion farming. Since 2002, the International Federation of Organic Farming Movements (IFOAM) has regulated total copper input on organic farms to a maximum of 8 kg ha^{-1} y^{-1}. These restrictions acknowledge the potential for copper levels in orchard topsoil to accumulate with repeated application. Have demonstrated that copper residues originating from fungicide application reduce soil microbial biomass while stressing the microorganisms. Therefore, it is necessary to better evaluate the potential impacts of copper contamination in agricultural land. In order to reduce the bioavailability of existing residues, management strategies and technologies need to be developed. On ferralitic soils of the tropics and subtropics, mineralization occurs much faster than on soils typical of temperate and continental zones. Consequently, a high organic matter content and high biological activity are essential for sustainable soil fertility. The beneficial impacts of organic farming on biological activity, microorganisms and soil organic matter content are, therefore, particularly important for soils in the tropics and subtropics.

Soil Structure

The environmental importance of a favorable soil structure lies in an improved resistance to structural soil damage, such as compaction and erosion. Soil structure can be measured by several physical parameters, such as the stability of aggregates, coarse pores, air capacity and water holding capacity. The maintenance of a favorable soil structure is a major concern in organic farming systems. Favorable rooting conditions as a result of microbial activity and a good exchange of water and air ensure improved spatial and chemical availability of nutrients. Evidence of increased aggregate stability under organic farming. Significant differences in soil structure when 16 fields of biodynamic or conventional commercial farms were compared in a paired study in New Zealand. There were also highly significant differences in total topsoil C and a range of physical parameters (e.g., reduced bulk density and penetration resistance and increased topsoil depth under organic and/or biodynamic farming). A similar paired study on a conventional and an organic farm in the USA and again found improved physical properties under the organic system. In contrast, found that in most relevant long-term trials in Europe no significant differences in soil physical parameters, like macropore volume, bulk density and soil stability, could be detected between organic and conventional farming systems. Again, a long time of organic management and longer-term changes may be more important than short-term effects.

Erosion

Soil erosion by wind and water is assumed to be the main cause of soil degradation worldwide. The loss of fertile topsoil by erosion results in a lower yield capacity, and in an undesired transfer of nutrients, pesticides and sediments in surface water. Although erosion partly depends on site-specific risk factors, such as topography and climate, the extent of damage by soil erosion can be limited by farm management practices. The C-factor (tillage and coverage factor) describes soil losses on a slope relative to soil losses at full fallow as figured in the Universal Soil Loss Equation (USLE). Organic farming systems are usually characterized by a lower C-factor and a reduced erosion risk because of the wider crop rotations. In addition, soil management techniques like organic fertilization, mulching and cover cropping improve soil structure and, therefore, increase the water infiltration and retention capacity, and thus reducing the erosion risk substantially. These management techniques are particularly relevant on the porous ferralitic soils of the tropics and subtropics to reduce the soil erosion risk after heavy rainfall. However, highly effective soil erosion minimizing measures like direct drilling and mulch drilling can be found more often on conventional farms than on organic farms as these measures require a herbicide based management. Comparative data for erosion under organic and conventional systems are rare. The organically farmed soil did not only have thicker topsoil but also had significantly higher organic matter content and less soil erosion than the conventionally farmed soil. In the long-term DOK-trial carried out by the Swiss FiBL relevant soil parameters of conventionally and organically farmed soils were compared. One of the results was that organic soil management improved soil structure by increasing soil activity, thus reducing the risk of erosion.

In conventional farming in the tropics, even flat soil gets eroded as a result of the use of herbicides and the lack of soil cover; however, organic farming can counter erosion successfully. In trials on a Cuban citrus plantation, the Cuban Citrus Institute and Swiss FiBL used locally adapted leguminous crops and were able to restore degraded soils very quickly, to successfully suppress weeds, to fix nitrogen and prevent erosion. The shift to biodynamic methods in the Ambootia Tea Estate (Dar- jeeling, India) has significantly reduced problems of soil erosion and risk of landslide, and increased retention of soil moisture in an area with very seasonal rainfall.

Desertification

Since organic farming techniques have the potential to improve soil fertility, soil structure and moisture retention capacity, organic management provides solutions to the problems associated with desertification (degradation of dry lands). Relevant techniques in this context include composting, mulching, use of cover crops, intercropping, the use of supplemental organic fertilizers and mineral fertilizers (e.g. rock powder, rock phosphate, potassium sulfate). Additional benefits are provided by the frequent use of endemic species which are more adapted to climate stress, as well as the use of water preserving and agro forestry techniques. With a high level of organic matter and permanent soil cover, the water and nutrient retention capacity is increased and microorganisms create a stable soil structure. Organically managed soils, therefore, tend to be more resilient to water stress and to nutrient loss. The high moisture retention capacity can substantially reduce the amount of water needed for irrigation. The organic farming's potential for countering desertification is demonstrated by some practical examples in arid areas. In Kenya, the International Centre for Research in Agro forestry (ICRAF) runs organic farming projects to fight drought. Agro forestry systems generate multiple benefits, including erosion control and moisture retention. The organic farm is situated in an arid area severely affected by Stalinization resulting from inappropriate irrigation techniques. It is a certified organic farm with more than 3000 ha of which 650 ha are cropped with mango trees. Conventional agriculture in the area uses the water from the rivers and causes further stalinization by over irrigation. By using only rainwater, the avoids the depletion of water resources and reduces soil Stalinization considerably. Fertilization regimes include grazing cattle under mango trees as well as application of composted manure. Organic farming offers a favorable option in arid areas, but has to face constraints such as a lack of knowledge, lack of organic materials, land-holding constraints and the perception of organic farming as being old fashioned. The examples above have shown the potential benefits of organic farming for countries hit by desertification. Organic management could be a key to bringing degraded land back into production and therefore significantly contribute to the solutions of the world food problem. Governments would need, however, to actively promote organic farming systems and land reform.

Ground and Surface Water

The harmful effects of intensive agriculture on ground and surface water are caused mainly by the following practices:

- Excessive application of mineral N fertilizers, nutrient surpluses and a high level of available nitrogen after harvest;
- High organic fertilization level in combination with high stocking rates;
- Contamination of water with synthetic pesticides; and
- Lack of a protective soil cover, a narrow crop rotation and frequent tillage.

In areas where water is scarce (e.g., in the Mediterranean countries and in arid and semiarid zones), excessive water use, particularly for irrigation, is a problem. National standards for organic farming in Europe set up limits for irrigation in order to conserve water resources.

Nutrient use and Balance

The general conclusion that can be drawn from the literature is that organic farms often have smaller nutrient surpluses than conventional farms. Obviously, the same farm types need to be compared because livestock farms often have higher nutrient surpluses. It follows that in organic farming, the risk of water and air contamination as a consequence of nutrient surpluses tends to be lower. Restrictions in organic standards include a ban of mineral N fertilizers and a limitation of livestock density. More characteristic of organic farms, therefore, is that N tends to be a minimal factor, particularly on arable farms. Since the opportunity costs to produce N on farm in organic systems can amount to from seven to 16 times the cost of mineral N fertilizers, it is of particular economic interest to avoid N losses. As far as nutrient deficiencies are concerned, argue that the medium-term effects of an unbalanced nutrient supply are likely to take the form of a reduction in economic performance rather than environmental detriment.

Nitrate Leaching

As a result of fertilizer or manure applications as well as N fixation by leguminous crops, N accumulates in the soil. Nitrate leaching occurs when the amount of nitrate in the soil exceeds the plants requirements and when water from rain, irrigation or snowmelt moves through the soil into the groundwater. Nitrate in water can lead to toxic contamination of drinking water for humans and animals, as well as a eutrophication with excessive algal growth. The most common parameters used are the nitrate leaching rate and the potential for nitrate leaching. Variation in leaching from individual fields is large both in organic and conventional agriculture. Many organic systems operate at a lower level of N intensity than conventional systems because of lower stocking rates and fertilization levels. Another reason for lower N losses in organic farming is that their application is bound to organic manure and its corporation into the soil. Straw based manure, common in organic farming, reduces the nutrient availability and the risk from runoff in comparison to slurry. Other organic farming practices which minimize losses are wide crop rotations, soil cover during winter, intercrops, undersides and fallows of several years. Nitrate leaching in meadows, where herbage is removed from the field, are generally small. Greater losses occur where pastures are grazed because of the large returns of N in excreta. The flush of N mineralization following the sloughing up of leys is another feature of organic systems that possibly increases the risk of nitrate leaching. Leaching from arable land is increased where fertilizer rates exceed crop requirements. In particular, losses are associated with the temporary nature of annual crops and, sometimes, the lack of synchrony between release of N from organic matter and crop uptake. Improving the fertility of organically farmed soils by building up the content of SOM and incorporating organic residues and manures may increase this risk.

Taking all these factors into account, overall leaching losses from organic farms tend to be less than from conventional farms. Using a modeling approach, found that conventional dairy farms in, New Zealand had higher annual losses than organic dairy farms. show that nitrate leaching rates in organic farming in most studies are significantly lower than those of conventional systems. However, if the nitrate leaching rate is related to the output of grain and milk, organic systems tend to perform similarly or even worse. The farm level data are supported by the results of nitrate screening. Large-scale surveys in water protection areas in Germany and Denmark indicate that organic farming results in a lower or at least similar potential for nitrate leaching into ground and surface water. The absolute values generally did not exceed critical levels. As a result of improved conventional (integrated) management of mineral N fertilization, or systems using intensification measures especially in water reclamation areas with strict regulatory standards and extensive control measures, differences between organic and conventional systems have become smaller recently.

Even though scientific results from other climatic zones are scarce, positive effects of organic farming on the nitrate leaching risk can be reported from a citrus farm in Cuba. Under organic fertilization management based on composting with 60 kg N ha^{-1}, the farm achieved exactly the same yield level as under conventional fertilization management with 200 kg of mineral N. This example shows that organic fertilization management can help reduce the risk of nitrate leaching, especially under extreme climatic conditions. Although not all the results showed that organic farming results in less nitrate leaching than conventional farming, a strong tendency towards a decreased risk of nitrate leaching can be deduced. Losses after ploughing of leys for instance, can be large. But the growing consciousness of problematic phases in crop growth or cultivation has resulted in improvements in organic management practices. The remaining differences may decline as conventional fertilizer practices improve under increasing regulatory pressure.

Phosphorus

The main pathway for P losses is by movement of soil particles (i.e., together with soil erosion). Leaching is a much smaller and a more site-limited effect. Although the quantities of P lost from farmland are usually small in agricultural terms, losses of a few kilograms of P per hectare are sufficient to be of environmental concern. Leaching is most likely on deep sandy soils or high organic matter soils with little capacity to absorb P. Data on P leaching and runoff from Organic Farming are scarce. As nutrient balances for organic farms rarely show a significant surplus of P, losses are assumed to be small. A more reliable indicator to determine losses could be the differences in the dominant loss pathways in livestock and arable farming. Because of the relatively infrequent use of P and potassium (K) fertilizers in organic farming (e.g., rock phosphate, sugar beet processing waste), organic systems have been criticized for exploiting reserves of P and K built up by conventional farming. Some authors have reported a decline in the concentration of extractable P and K in soils after conversion to organic management, but this decline does not always occur, even where budget deficits of P and K are measured. In soil fertility in organic farming systems in England. Their results support the argument that organic arable systems are mining reserves of P and K. Therefore, they state that changes to organic management practices are required in order to increase inputs of P and K, if long-term declines in soil fertility are to be avoided. In the case of broad acre organic farming in Australia, soil available P has declined to levels where yields are being compromised and the sustainability of the farming system is being questioned.

Pesticides

The term pesticide covers a wide range of chemicals (e.g., fungicides, herbicides, insecticides, acaricides, algicides, lumbricides, molluscicides, nematicides, rodenticides, and plant growth regulators). Pesticide use in organic farming is very restricted. Only a few pesticides are approved for organic use (e.g., sulfur, copper, natural pyrethroids) and tend to be used only as a last resort for minority or protected crops. Synthetic pesticides are completely banned. In addition to the pollution of surface and groundwater, there is also a risk of air and soil contamination. Many reviews come to the same conclusion: because synthetic pesticides are not permitted for use in Organic Farming, the risk of contaminations of air, soil and water in this respect is avoided. The spraying of pesticides permitted in organic farming is connected with a comparatively negligible risk due to low volatility. The exposure of certified biocides is measured at extremely low levels when compared to conventional systems in organic permanent crops, these being more prone to pests and diseases. The application of powdered and fluid substances permitted by organic standards may cause a short-time impairment of air quality. The impact of pesticides on water quality in organic systems has rarely been studied. Again, however, most of the water contamination comes from herbicides used in conventional farming. There is some debate about the disposal of sheep dip and the relative risks of pyrethroids versus organophosphates. Organic farmers only use the former and they are potentially more damaging to aquatic habitats.

Pathogens

Pathogenic organisms from livestock can contaminate surface waters used for drinking, bathing or irrigation. The application of organic manures to agricultural land is one route by which pathogens may be introduced into the human food chain. There are only few data available on the relative risks of pathogen transfer from organic and conventional farming systems. Pathogen levels can decline during manure storage, particularly if solid manure is actively composted to increase the temperature of the heap. Thus, it might be concluded that biodynamic farming provides a lesser risk because manures are more often composted. So far, there are no data to prove this, but research is on-going. The overuse of antibiotics and pesticides in conventional farming, in horticulture and especially intensive enterprises like poultry farming, has the potential to lead to drug resistance of pathogens and consequently to the contamination of the environment and the food supply. In organic farming, the use of regular synthetic veterinary medicines is restricted and the use of antibiotics for preventive treatments prohibited. Therefore, Organic Farming does not carry the same risk as conventional agriculture.

Climate and Air

Global climate change (greenhouse effect) is considered one of the most urgent environmental problems. The gases carbon dioxide (CO_2), nitrous oxide (N_2O) and methane (CH_4) mainly contribute to the greenhouse effect and are largely, directly or indirectly, as a result of the burning of non-renewable resources. On a global scale, agriculture is responsible for roughly 15% of the trace gas emissions with climatic impact. However, agriculture also provides a sink for CO_2 because of the fixation of carbon by crops and pasture. An important side-effect of global warming could be that cultivation zones will shift pole wards and that agricultural yields will be affected. Extreme climatic events will occur more frequently and that this will jeopardize plant production.

Carbon dioxide

Carbon dioxide emissions from the agricultural sector in OECD countries are estimated at less than 1% of overall CO_2 emissions (IPCC 2001). Net emissions of CO_2 from agriculture depend upon the direct and indirect use of fossil fuels, and on the amount of carbon sequestration in soil organic matter and crop growth. In order to compare farming systems CO_2 emissions, need to be differentiated between the emission due to the burning of fuel (direct energy) and the fuel used for the production and transport of fertilizers, machinery and synthetic pesticides (indirect energy). 70% of CO_2 in organic farming resulted from fuel consumption and the production of machinery, whereas 75% of the CO_2 emissions in conventional systems were due to N fertilizers, feedstuff and fuels. On a per hectare scale, most studies found lower (up to 40-60%) CO_2 emissions in organic systems. The main reasons for these positive effects are the renouncement of the use of mineral N fertilizers with high energy consumption, lower use of high energy consuming feedstuffs and mineral fertilizers as well as the elimination of pesticides. But, on a per unit output scale, which mainly depends on the yield that is achieved, CO_2 emissions tend to be higher in organic farming.

Carbon Sequestration

Soil carbon levels have decreased under agricultural land use. Therefore, sustainable agricultural strategies including recycling of organic matter, tightening nutrient cycles, and low tillage or no-tillage practices may rebuild organic matter and reduce losses from the system. Generally lower crop yields, plant productivity in organic farming accounts for almost the same organic matter return as in conventional systems. Over a 15-year period that mixed farming with manure application and the combination with other organic farming techniques lead to significantly higher organic matter levels in soil as compared to conventional farming. There is a huge potential for CO_2 sequestration, which differs for tropical and temperate countries. Developing countries, mostly in the tropical belt, have a 30-60% higher potential than industrial countries, mainly located in temperate climatic regions. Agro forestry holds the highest potential of agricultural carbon sequestration in tropical countries and is seen as a viable alternative to slash-and-burn agriculture in the humid tropics. Simple agro forestry systems with one species (e.g., oil palm, rubber) can regain 35% of the original carbon stock of the forest, which is three times more than cropland and pastures. Extensive research on the situation and on the potential impact of agro forestry has been undertaken by ICRAF for East and Southern Africa. Carbon stocks can be tripled over 25 years, similar results can be assumed for sub humid West Africa and sub humid South America, but there is little researched evidence. Sound technologies have been developed to sequester additional carbon in cropland management and in agro forestry and they are systematically applied and improved in organic farming. The evidence of their performance is still fragmentary, however, and does not allow regular comparison and quantification for the various agro climatic regimes and socioeconomic patterns. Organic farming enables ecosystems to better adjust to the effects of climate change and it also offers a major potential to reduce the emissions of agricultural greenhouse gases. Moreover, mixed farming and the diversity of organic crop rotations are protecting the fragile soil surface and may even counteract climate change by restoring the organic matter content. The carbon sink idea of the Kyoto protocol may therefore be accomplished efficiently by farming organically.

Nitrous oxide

Nitrous oxide contributes severely to global warming and the depletion of ozone in the stratosphere. Almost 90% of the global atmospheric N_2O is formed during the microbial transformation of nitrate (NO_3^-) and ammonia ($NH4^+$) in soils and water. In OECD countries the agricultural contribution to N_2O emissions is estimated at 58% (IPCC 2001). Soils fertilized with inorganic fertilizers and manure stores are seen as the largest sources. Nitrous oxide emissions are very difficult to measure and, therefore, have been related to the total N input in the form of fertilizers, manures and crop residues. Consequently, it has been largely assumed that, because organic farming operates at a much lower intensity, with lower N inputs and less available mineral N in both manures and soils, N_2O losses will also be lower. However, until recently there have been no quantitative comparisons between organic and conventional systems. Within conventional agriculture, the main risks arise from manures and from the water logging of soils by heavy rainfall following fertilizer application. Within organic farming, the main risks come from manures and the incorporation of residues from legumes. In the absence of direct measurement, one may assume that the amount of N_2O lost per unit of yield is unlikely to differ to that from conventional systems, but losses per unit area may differ, depending on the cropping system and input of organic manures.

Methane

Calculates the CH_4 share in the greenhouse effect of about 2.5% and agriculture is believed to account for roughly two-thirds of the total human-generated CH_4. While paddy rice fields, cattle feedlots and the burning of biomass contribute to methane emissions, about 75% of methane on farms is emitted directly from ruminant animals, from digestive processes and excretion. In order to assess the overall methane emissions from farming systems, several factors like animal numbers and type, diet and manure management system need to be considered as well as their interactions. Diets, that are high in roughage, for example, will release higher rates of methane than diets high in starch. This may result in higher emissions from organic systems, where diets tend to be high in roughage and low in concentrates. The higher proportion and lower productivity of ruminants in organic farming may lead to slightly higher CH_4 emissions. Standards and breeding programs in organic systems aim at longevity in order to prolong the productive period in relation to the unproductive life of young cattle. Correspondingly, the unproductive CH_4 emission of calves and heifers may be reduced. Comparative empirical studies on CH_4 emissions in different farming systems are scarce. Compared a conventional and an organic farm rearing beef cattle in southern Germany and calculated that CH_4 emissions were about 25% higher on the conventional farm. On the basis of the literature reviewed and expert knowledge, only the following conclusions (based mainly on studies for milk production) can be drawn: as a result of lower stocking densities, organic farming might have a lower CH_4 emission potential on a per hectare scale, whereas per unit output, the CH_4 emission potential tends to be higher than in conventional farming. However, in the absence of extensive solid data, no significant differences between the two farming systems with respect to CH_4 emissions can be identified.

Ammonia

Ammonia does not contribute to the greenhouse effect, but causes acidification and eutrophication when redeposit to soils and water and can damage sensitive habitats. In agriculture, livestock production, in particular, accounts for the main part of NH_3 emissions. Ammonia is produced when urea in urine and dung comes into contact with the enzyme urease, which can be found in both manure and soil. Therefore, animal housing, manure stores and the spreading of manures to land are major sources of NH_3. There has been a large amount of research into NH_3 emissions from conventional animal production but only a few studies specifically on organic farms. Differences in dietary N intake and N excretion, housing system and period, manure storage and spreading, and livestock density will affect the amount volatilized. Since organic systems operate at a lower level of intensity, NH_3 losses may be lower too. Emissions from housed animals are considered to be greater than those from grazed, as urine is quickly absorbed into soils. In organic systems, maximum grazing is recommended, which tends to reduce housing periods. Therefore, the potential for ammonia loss is likely to be less, although this has not been tested. Straw based systems also tend to have lower emission rates than systems based on slurry; the latter are more common in conventional farming. Furthermore, much evidence suggests that ammonia losses are greater from composted manures compared to those which are just stockpiled.

Organic farming tends to bear a lower potential for NH_3 emissions than conventional farming systems. In a scenario of complete conversion to organic farming, estimated a potential reduction of acidification by more than 30%, caused mainly by reduced ammonia emissions. In contrast, provides a risk assessment upon which NH_3 emissions will not necessarily be lower in organic farming than in conventional. In the absence of direct measurements, one may assume that the amount of NH_3 lost per unit of yield is unlikely to differ to that from conventional systems, but that losses per unit area are likely to be less, due to lower livestock densities.

Energy

The OECD (1997) proposed to use energy intensity and efficiency as appropriate indicators to measure and evaluate energy use. The corresponding parameters are:

- Energy consumption (per hectare and per output); and
- Energy efficiency (input/output ratio).

Several studies have been conducted on energy use in organic farming systems that differ in methodology and in the boundaries of the systems being studied. Furthermore, there is no clearly defined system of conventional farming, which ranges from high input intensive systems to near-organic systems. Therefore, it is difficult to make meaningful comparisons. This must be taken into account when drawing conclusions from studies that compare organic and conventional systems.

Intensity of Energy use

Direct on-farm consumption of fossil energy (e.g., for fuel and oil) needs to be distinguished from indirect energy consumption, which results from the production of synthetic fertilizers and pesticides, the transport of imported feedstuffs and from investment goods such as agricultural machinery. Inputs of direct energy per unit area in the long term DOK trial in Switzerland were similar across conventional, low input and organic systems. Since basic operations such as ploughing, cultivation, sowing and harvesting are likely to be similar, reduced fuel costs in organic systems due to the absence of most pesticide applications and lower harvesting energy inputs because of lower yields, are more or less balanced by increased fuel use for mechanical weed control. Inputs of indirect energy tend to be substantially lower in organic systems. The major difference is the greater energy use in conventional systems to produce and transport fertilizer, particularly N fertilizers.

If both direct and indirect energy uses are considered together, calculations of energy consumption per hectare indicate that organic farms use less energy than conventional farms. Calculated that average energy consumption on organic farms amounts to 64% of conventional farms. The energy consumption of organic farms amount to 30% to 50% of conventional farms. For organic potatoes and apples, energy consumption per output unit is higher relative to conventional production. This is the result of a higher energy input for mechanical measures like weed control and the lower mineral N fertilizer use in conventional production. Lower energy consumption on organic farms for olive and citrus production in Sicily (Italy), both with regard to energy consumption per hectare and per output unit. A comparison of the energy use per hectare for organic and conventional farming (e.g. in UK, USA, the Philippines) showed a 30% to 70% lower consumption per unit of land for organic systems.

Efficiency of Energy use

There are varying results on the energy efficiency of different farming systems. In many organic systems, the yield of crop and animal products is less than in conventional systems. However, the size of the differences will depend on factors such as farm type, soil type, climate and the intensity of production. Furthermore, no standardized scheme for calculating energy use efficiency exists. State that analysis should take account of overall farm energy balance and also include activities that are not crop specific (e.g., handling and application of manures and fertilizers, winter catch crops, use of fallows for weed control). However, apart from a few studies based on long-term rotation experiments these energy inputs have not been considered. Thus, comparing individual research results in this context is only of limited value. Little difference in overall energy efficiency when they compared mixed sheep and arable farms in New Zealand. But they noted that the conventional farms relied more on legumes for N supply, and therefore were more energy efficient than European equivalents.

Compared conventional, minimum tillage and alternative (equivalent to organic) systems of growing soya, wheat and barley over seven years in South Dakota, USA. Overall, the alternative system had the greatest energy efficiency. The minimum tillage system had the lowest efficiency because reduced direct energy input, as tractor fuel, was more than balanced by increased fertilizer and herbicide energy input. Comparing rotations of different productions systems in Iran, found the energy efficiency of organic farming to be 81% better compared to high-input conventional farming. In a similar investigation in Poland, calculated a 35% higher energy efficiency of organic compared to conventional farming. Overall, the literature review suggests that organic methods generally use less energy per unit area and per unit of output, both for individual crops and livestock types, as well as on a whole farm basis. Again, however, the setting of system boundaries, methods of calculating the energy values of inputs and methods of calculating energy use efficiencies vary substantially between studies. Comparisons across studies are hardly possible there is an urgent need for a commonly agreed standard methodology.

The evidence presented in this chapter indicates that there is wide agreement that organic farming comes closest to an environmentally friendly agriculture (Table 4.2). Particularly pronounced is the significant level of pesticide pollution in conventional agriculture; in organic farming, in contrast, there is only a very limited use of pesticides. The differences are likely to hold whether assessed per area or per unit of food produced. A second major area where organic farming is more environmentally friendly is soil conservation. Soil care is a guiding principle in Organic Farming. It is expressed in higher levels of soil organic matter, the active promotion of soil biological activity, more balanced nutrient cycles and lower soil erosion risks. A third main benefit is the expressed goal to enhance biodiversity: organic farming depends upon intact ecological balances and favorable biological processes expressed in ecological services like pollination or pest control by natural predators and others conclude that organic farming creates more favorable conditions at the species and ecosystems level of floral and faunal diversity than conventional farming systems. Less affirmative, though not necessarily less favorable than for conventional systems, is the evidence that has been presented in fields like pollution of water resources and the food chain with pathogens (due to the more pronounced use of organic fertilizers and manure). The same applies to the emission of N_2O and CH_4 (because manure stores are seen as a major source and because, on an output unit scale, the CH_4 emission potential tends to be higher in organic farming). Organic Farming is not exempt from the trend of intensification. Therefore, others stress that the increasing economic pressures endanger the positive aspects of organic farming. In order to maintain the view of nature conservation of valuable grassland, for example, a very low intensity of production is necessary. But because of increasing intensification, this is hard to realize.

Table 4.2: Overview of the absolute and relative environmental impacts of organic farming compared with conventional farming (author's compilation based on the empirical material presented in this chapter)

Area	Aspect	Environmental impact[A]	
		Absolute	Relative
Biodiversity	Genetic diversity	+	+
	Floral diversity	+	++
	Faunal diversity	+	+++
	Habitat diversity	+	+
Landscape	Landscape structures and aesthetic value	+	+
Soil	Soil organic matter and acidity	+	++
	Biological activity	+	+++
	Soil structure	+	+
	Erosion	–	++/–
	Desertification	+	+
Ground and surface water	Nutrient use and balance	–	++
	Nitrate leaching	–	++/–
	Phosphorus	0	+?
	Pesticides	–	+++
	Pathogens	–	–
Climate and air	Carbon dioxide (CO_2)	+	+
	Nitrous oxide (N_2O)	–	+/-
	Methane (CH_4)	–	–
	Ammonia (NH_3)	–	+/–
Energy	Intensity of energy use	NA	++/–
	Efficiency of energy use	NA	+

[A] + = Slightly better; ++ = better; ++ = substantially better; ++/– = better with some aspects that are negative; +? = better with some uncertainties; +/– = partly betterand partly worse with some uncertainties; = unclear; – = negative impact; 0 = no impact or change; na = not applicable.

The same economic pressures lead to a rationalization of organic production that, in many respects, starts to resemble the rationalization path of conventional agriculture. Field parcel sizes are being continuously increased, agricultural workers are being replaced by larger machinery, and livestock husbandry systems are becoming more rationalized and often more intensive. Clearly, organic farming has the potential to develop sustainable land use systems, but the motivation of farmers and financial support are the main factors required to achieve this goal. Organic management could also be a key to bringing degraded land back into production and therefore significantly contribute to the solutions of the world food problem.

Organic Farming in Protected Areas

Protected areas are dedicated to the protection and maintenance of biological diversity (as defined by the World Commission on Protected Areas). Since organic farming can support biodiversity, it offers an important agricultural management option. Organic farming is specifically promoted in protected high nature value areas. Not only in developed countries but also in the developing world the nature conservation agenda is one of the driving forces behind the growth of organic farming. A recent literature survey highlights examples where nature conservation organizations are working closely with local farmers who live in, or close to, areas of significant nature conservation interest. Organic farming practices are promoted. In the Meso-American Biological Corridor, a projected complex of protected areas and sustainable management stretching over seven countries, a range of sustainable land uses within buffer zones and linking areas have been explored, including certified forest management and Organic Farming.

Future Research needed

The main gaps in the literature (and presumably research) on the environmental effects of organic farming can be summarized as follows.

- Scientific evidence of the environmental impacts of organic farming for the Southern hemisphere is rare. More research on the environmental services and benefits or impacts of organic farming is urgently needed.
- The knowledge of the impacts of organic farming in pastoral and upland agriculture is limited. Therefore, a need for longitudinal, system-level studies that address these issues remains. Furthermore, well-replicated studies are required that follow the development of flora and fauna on organic farms during and after conversion, and compare them with neighboring farms that continue with conventional management.
- The complexity of interactions between species and ecosystem functions are not yet fully explored so that simple biological indicators of soil productivity are yet to be identified. In particular, more research is needed on the interactions between the diversity and density of habitats in the agricultural landscape, and the performance of production systems. Both the positive and negative interactions between cultivated and uncultivated areas in organic farming systems need to be covered. Reliable indicators of a supposed functional integrity need to be developed.
- Comparatively few data are available on the effects of organic farming on wild animal groups (e.g. birds, mammals, amphibians, arthropods). There is particular urgency with regard to the questions of pollinators.

Organic farming (like conventional farming) is not able to support highly specialized species which depend on complex mosaics of habitats. Strategies to optimize nature conservation within organic systems are therefore required. Particular attention should be given to developing organic farming practices that support the healthy functioning of the relevant ecosystem processes and functions.

- The basic principles of organic farming provide opportunities for the further development of high-quality landscapes and a cautious utilization of sensitive areas. How precisely organic farms (can) contribute to landscape quality should become a main area of further research. There is, for example, little information available to compare habitat diversity in organic and conventional farming systems. Particular attention should be given to developing reliable indicators of both the positive and negative interactions between cultivated and uncultivated areas in organic farming systems.
- Only very scarce research is available on the interrelationships with climate (change) and air pollution. Studies are needed which analyze CO_2 emissions and accumulations of different farming systems in a net balance approach.
- The development of methodologies for evaluating landscape quality has just begun and quantitative research investigating the impacts of different farming systems on landscapes is scarce. Clearly, the development of methodologies that deal with landscape quality in an integral way deserves greater attention in order to assist the evaluation of environmental impacts. Particularly little is known about the influence of Organic Farming to aesthetical attributes of the farming landscape. At the same time, the assessment of landscape in terms of an areas visual character is likely to become increasingly important as public money is used to support delivery of social goods. How an organic farm can contribute to the landscape should be a main area of interest for further indicator development.
- There is an urgent need for a commonly agreed standard methodology for calculating energy use efficiency. Critical questions include the setting of system boundaries, the methods used for calculating the energy values of inputs and the methods used for calculating energy use efficiencies.
- As for the development of agri environmental indicators, there has been substantial progress since 1990. Particularly useful are the schemes developed by OECD, Euro state and the European Environmental Agency (EEA). The main challenge is to apply these schemes in practice, because in many areas and in most parts of the world, an environmental monitoring that is sufficiently disaggregated in temporal and spatial terms is still rudimentary.

5

A Plant Agronomy Study

Organic crop husbandry is based on practices outlined in the various organic standards that have been developed and documented in many countries. The most widely recognized organic standards are those published by the International Federation of Organic Farming Movements (IFOAM 2002). Organic farming methods emphasize the use of internal on-farm inputs rather than externally sourced inputs to achieve essential soil fertility, nutrient management and plant protection goals. Self-regulation within an agro ecosystem, multi-year management cycles and a focus on prevention rather than reaction, are key principles in organic farming that underpin organic plant production. The core element in organic crop husbandry is the management of a site-specific and market-oriented crop rotation. Alternating a wide range of crops over time and space can facilitate efficient use of the soil resource of a farm by cultivating crops of varying profiles in nutrient demand and supply, growth habit and phytosanitary characteristics. As the specialization in stockless farming expands regionally and globally, the importance of crop rotations in nutrient management increases and there is greater reliance on off-farm sources of manures. While the trend towards stockless systems may raise questions about sustainability in some areas, the more crop husbandry is market-oriented or commercially driven, the more conflicts are likely to arise as good organic agricultural practices are compromised. For example, the need to forego sufficient soil fertility building components in a rotation, whether with livestock or legumes, in favor of more components as cash crops.

In addition to crop rotations, a range of other cultural strategies used in organic farming to achieve various farm management objectives are discussed in this chapter. Organic Farming is an integral part of broader landscape management. Even small-scale farming can contribute to increase and enhance the proportion of non-productive area of a holding. In some cases, especially on leased farms, the establishment of perennial components such as hedges, tree lots or ponds is difficult to implement. As an alternative, flowering field margins or corridors can be integrated as an annual enrichment of the agro ecosystem, increasing the abundance of faunal elements for better self-regulation among adjacent fields. The main focus of this chapter is mixed and stockless annual cropping systems, rather than purely perennial farming systems such as tree and vine crops. The success of organic crop husbandry also relies very much on effective weed management, although diverging views can be found regarding the necessity of clean crops versus the approach of higher agro biodiversity through more natural, wild flora components in a field. Weeds are widely reported as a major constraint in organic production and severe crop losses can be experienced, particularly during conversion and in phase changes in rotations (e.g., from pasture to arable cropping). A review of plant pest and diseases in organic farming is presented in the following chapter.

While many important generalizations can be made about soil processes, plant growth and so on, the performance of particular farming systems, cropping rotations and crop varieties will always be determined by the local conditions. Historically, most of the research into organic farming has been conducted in Western Europe, the home of modern Organic Farming. However, when the environmental and market conditions are different to those experienced in Western Europe, organic farmers need to be innovative and incorporate local knowledge.

The central role of crop rotations in organic farming is highlighted by the important focus on rotations in many organics' certification standards. For example, in the United Kingdom (UK), the standards for plant production (including pastures) require a multi-annual rotation programme for building soil fertility and managing weeds, pests and diseases. Crop rotation is the term used to that crop are grown over time in a very specific order. After a number of years (length of the crop rotation), the cycle will be repeated. The crops grown in one year on the available area of a farm make up the cropping plan. If the crop rotation is consistent and unchanged, the cropping plan is the same every year. Crop rotation has a temporal aspect [where] crops are grown over time in a specific order (succession of crops in time); and a spatial aspect [where] the crops grown this year and their division over the available space. The interaction between spatial and temporal aspects can be used to strengthen the crop rotation concept. Rotating the crops on the available space is done so that a given crop is never grown next to a field with the same preceding crop (spatial crop rotation). This helps to prevent semi-mobile pests and diseases from surviving from one year to the next.

The aims of crop rotations in organic farming systems were summarized:

- Maintenance and improvement of soil fertility
- Maintenance and improvement of soil organic matter
- Maximization of symbiotic N fixation through cultivation of forage and grain legumes
- Production of sufficient food and straw for animal husbandry
- Optimized use of pre-crop effect through crops with high gross margin
- Mobilization of nutrients through crops with high root density and root depth
- Control and reduction of pests and diseases
- Control of weed competition combined with gentle soil cultivation
- Improvement of time management for crop husbandry-specific labour
- Maintenance and improvement of the economic situation of the farm.

In mixed farms, the crop rotation is mainly focused to provide enough fodder for the different animal species. Therefore, the rotational plan will react to market demands only to a limited extent. This can be contrasted with farms with little or no livestock, in which case commercially desirable crops may dominate the crop rotation. From the production point of view, the rotational system has to meet requirements for the maintenance of soil fertility, nutrient supply and control of weeds, pests and diseases. Economic aspects also influence decisions about crop rotations, such as labour management and maintaining the continuity of farm cash flow through fluctuating market and circumstances. In organic farming, the tension between short term (economic) demands and long-term (ecological) needs creates challenges for farmers, and complex skills are required to balance the competing interests by managing a sequence of crops that provide direct cash benefits as well as indirect ecological services.

Management of Nutrients

The structure of organic rotations generally consists of two parts. First, legumes are used as a soil fertility-increasing component, from yearly to multi-year crops, mainly in the form of forage legumes, with much less in the form of grain legumes and, second, periods with non- legumes such as cereals, root crops or field vegetables, relying on the accumulation of humus, organic nitrogen (N) and depleting the resources. In organic farming, the emphasis is on developing healthy, biologically active soil through the use of certified inputs and land management practices. Inputs used in rotations include plant residues, animal manures, rock dusts and biological activators (IFOAM 2002). Additionally, biological activity is promoted so that soil processes such as nutrient cycling and soil structural development are facilitated. Crop and pasture plants then rely on the mineral nutrients that are made available in the soil have indicated that the underlying soil processes and nutrient pools are similar in organic and conventional soils although the relative rates and importance of individual processes differ between farming systems. While many soil changes under organic management may be clearly significant, other research shows that changes are often subtle rather than dramatic.

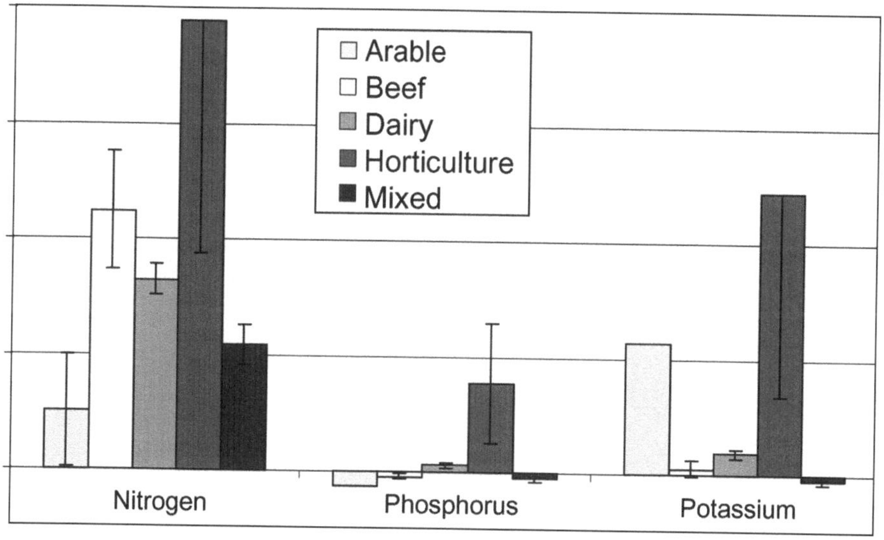

Fig. 5.1: Average farm-scale nutrient budget surpluses (average input minus output) for nitrogen, phosphorus and potassium on various organic farm types in temperate countries

Careful N management is needed for most crops in order to maximize yields while minimizing nutrient leakage. In organic farming, this mobile nutrient is usually not difficult to supply, whether through legumes or animal manures, but is easily lost from the system. Phosphorus (P) is a very immobile nutrient, especially at higher pH levels, and organically certified sources of P commonly have low solubility. As a consequence, several recent studies have highlighted the lack of sustainability in existing organic P management in different settings. Using farm-scale nutrient budgets, it is confirmed that N was generally not a problem but that P and K were being depleted in some systems, particularly arable cropping.

Dairy farms had low nutrient budgets as a result of the very low reliance on external inputs, while horticultural farms had noticeably high budgets, presumably as a result of large externally sourced inputs of manure. Where the older organic farms have less P than newer organic farms a trend not observed in paired conventional farms. By including greater carbon (C) inputs in the soil, organic producers are able to store nitrogen and other nutrients for future crop use, while minimizing the risk of environmental pollution. Asserted that improved soil structure was dependent on frequent, and presumably large, inputs of fresh organic matter, a common practice in Organic Farming. Review of several conventional field trials comparing the long-term (20-120 years) effects of fertilizers and manures (farmyard manure, slurry and green manure) on crop production and soil properties indicated that large amounts of manures and other organic inputs were needed to get improved soil chemical and physical properties, but crop yield was not affected.

Green Fertilizers

The cultivation of crops that are not devoted to selling or feeding purposes is commonly understood as green manuring. Older definitions focused more on the fertilizing aspects that were achieved by incorporation of crops in an immature stage.

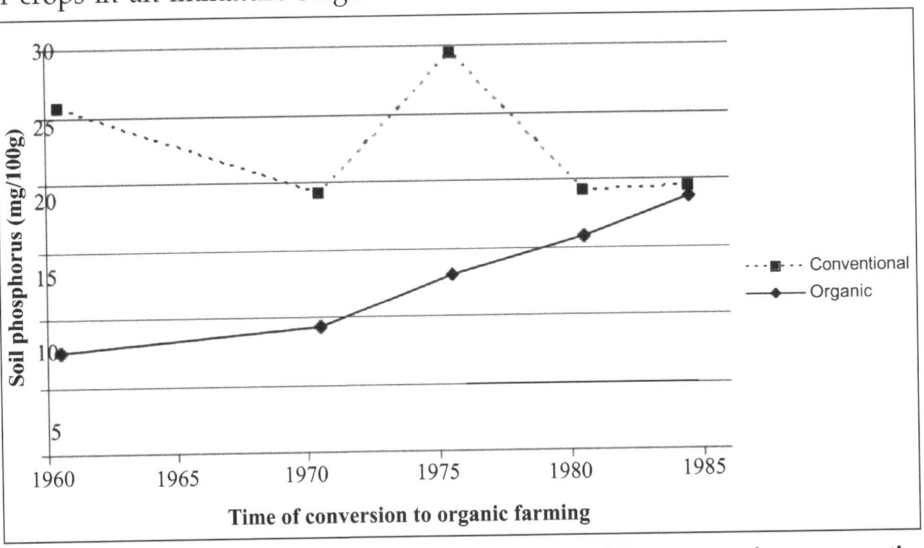

Fig. 5.2: Phosphorus concentration in the top 20cm of soil for German farms converting to organic practices in the years from before 1960 to 1984 compared with paired conventional farms

More recent understandings emphasize a variety of pre-crop effects including soil physical, chemical and biological aspects, as well as ecological services such as crop protection and weed control. Green manuring in a rotation contributes to a higher level of biodiversity in time and space.

Depending on the period and the purpose of cultivation, there is a distinction between green manure as main crop, intermediate crop and companion crop. Terms such as catch crop or cover crop are synonymous with intermediate crop. The first term clearly indicates the expectations for N-conserving activities in the crops used, mainly fast-growing crucifers and grass species. The second term is similar, but also emphasizes the soil conservation action of covering the soil with vegetation. Under sowing or, more recently introduced, living mulch systems increase the complexity of a growing system by combining different partners at the same time. Often, such an approach in planting design is also named the intercropping system, a concept that was commonly applied in traditional farming systems throughout the world. It now covers a wide range of techniques such as under sowing, dual cropping, strip cropping and even agro forestry, hedging, shelterbelts and windbreaks. All of these methods involve increased plant diversification within a farm and can result in increased opportunities for agro ecological processes to occur. The potential benefits of intercropping include improvements in soil fertility, resource capture, pest and disease control, weed management and risk management. However, intercropping requires good farm management skills to satisfy the full agronomic needs of more than one crop.

Intercropping systems using in situ mulch crops have been widely tested, particularly in the United States of America (USA). Using leguminous cover crops in cabbage found soil fertility and weed control benefits but also noted that interspecific competition was a major problem. Careful matching of secondary crops with the primary crop and the prevailing growing conditions is needed based on relative vigour, timing of growth phases and complementarily of resource use. Intercropping may be especially suitable for row crops with poor competitiveness against weeds.

Perennial farming systems with tree, bush or vine crops also provide niches for short- and longer-term intercropping. In addition to the use of cover crops on the orchard floor of many crops including apples and wine grapes, plantings of commercially and environmentally valuable species within or around the orchard have also been used for pollination, disease and pest control in crops such as such as leeks and carrots. Within the European Union (EU), specific programs facilitate the use of one-year set-aside crop mixtures. This is often managed within a biennial or longer grass clover crop, out of which one year is financed by subsidies and not fed to animals. Similar effects, although on a lower level, can be expected from a one-year forage crop by excluding the last cut as roughage and providing additional biomass as C and organic N supplied to the soil. The one- year green manuring is most valuable as a humus source, whereas the quickly degraded biomass of green manure provides more nutrients and energy sources for soil organisms and improves the fertility for the subsequent crop. Forage legumes were found to be more efficient in supplying N to subsequent wheat, crops than grain legumes such as pea (Pisum spp.) or vetch (Vicia spp.). The improvement in soil structure by the root residues of the preceding forage legumes was more important for the performance of a succeeding lettuce (Lactuca sativa) crop than the impact of the N supply alone.

Table 5.1: Types of crop rotations F, forage legumes; G, grain legumes; N, non-legumes; C, cover crops; U, under sowings

No. of fields	Year							F%	G%	C, U%
	1	2	3	4	5	6	7			
4	F F F	N N F	N+C GN	N+ UN +U N+U				25 25 50	0 25 0	50 25 25
5	F F F	N F F	N+ C N N+U	G N+C N+C	N+ UN +U N+U			20 40 40	20 0 0	40 40 60
6.	F F F	F F F	N N F	N+C N+C N	N+C GN	N+ UN +U N+U		33 33 50	0 17 0	50 33 17
7.	F F F F	N F F F	N+ C N N F	N+ CN C+ N+ C N	N N+ C N+ C N	G N N+C +C	N+ UN +U N+ U N+U	15 28 28 42	15 0 0 0	42 42 57 28

Allelopathic properties of glucosinolate-containing crucifers are highly relevant for their phytosanitary effects (biofumigation) against the incidence of soil borne diseases. Buckwheat as green manure may act similarly against various weeds as a result of the presence of gallic acid. Infestations by root knot nematodes, a severe pest in vegetable cultivation, can be reduced by various green manures such as radish, marigold and Sudan grass. Living mulches can be very effective against various pests as a result of increased predator occurrence or increased disorientation of herbivores. Whereas, very limited use of green manure crops as part of crop rotations in Dutch organic farms (3 of 68 interviewed farmers), German surveys in organic agricultural and vegetable holdings revealed that 60% of arable farms and 63% of vegetable farms used cover crops or under sowing. In a survey of organic farmers in the USA, the most common (72%) soil fertility management practice reported was the use of cover crops.

Generating Rotations

A range of rotations that differ in length and structure, specifically the proportion of forage legumes, grain legumes and cover or under sown crops. The grain legumes should not exceed a proportion of 33%. Forage legumes, if cultivated in mixtures of clover and/or alfalfa and grass species, can achieve mixtures up to 50%. Because of income possibilities, soil borne diseases and other factors, higher proportions of legumes can cause serious yield losses in a long-term perspective of rotations. The increase of species biodiversity in organic farming is exemplified by the range of potential cover crops and under sowings used in the rotations. Under favorable conditions forage legumes can be integrated into the rotation as under sowing in the preceding cereals. If the performance of the under sowing crop is poor, it should be replaced by a stubble crop sown after cereal harvest. However, the one or multi-year forage legume-part of the rotation is too important for the overall success of the rotation (e.g., for N fixation and other tasks) to be able to accept a poor performance. Cover crops may be best suited when winter crops are substituted by spring crops. The time-span between pre-crop harvest and subsequent crop sowing allows sufficient time for accurate seedbed preparation, intensive mechanical weeding if necessary, and substantial root development and above-ground biomass for soil protection.

In response to the needs of different farm types, the proportion of the various crop groups used on the rotation will vary slightly. Mixed farms with ruminants produce mainly rough fodder that can be part of the rotation of between 30 and 50% grass–clover or grass– alfalfa mixtures. Monogastric animals such as poultry and pigs are mainly fed by grains of legumes and cereals. Therefore, the proportions of legumes and cereals are shifting more to cereals and from forage to grain legumes. On arable farms the need for forage legumes is diminishing from the economical point of view. But from the perspective of sustainability, these farms should stay on at least a one-year set-aside system. Cooperation between mixed farms (preferably organic) could expand the forage legume segment from one to two years, enable the import of animal manure as a substitute for the produced and exported forage fodder, and could keep the farm arable with great benefits for labour management. These need to be adjusted for specific site and crop conditions. In more arid zones, spring cereals can be better suited than winter cereals. If harvest periods tend to be too late for the timely sowing of succeeding crops, early maturing cultivars have to be chosen for cultivation. Depending on soil moisture conditions, cover crops may be integrated into the rotational plan instead of partial fallows. Their value cannot be assessed only by narrow economic criteria. Indirect effects become more obvious over the longer term, including improvements in soil organisms, soil structure, soil management, traffic ability, infiltration rate, and mineralization rate.

Table 5.2: Examples for proportions (%) of typical crop groups (Freyer 2003)

Farm structure	Legumes	Cereals	Root crops	Cover crops
Mixed farm (milk cows)	30-50[A]	30-50	5-15	20-50
Mixed farm (various animals)	25-40[B]	40-60	10-20	20-50
Mixed farm (pigs)	20-35[C]	50-60	15-25	40-60
Arable farm	25-30[D]	40-60	20-30	40-60

[A] Mainly forage legumes; [B] forage legumes (>50%), grain legumes; [C] grain legumes, forage legumes, propagation of cover seeds, set-aside; [D] grain legumes, forage legumes, propagation of cover seeds, set-aside.

In regions where rainfall is lower and/or less reliable, such as in Western Australia, simpler rotations with longer pasture or fallow phases are more likely. If the soil is heavier, with better nutrient and water retention, a suitable organic crop rotation would be pasture (vetch or medic Medicago spp. hay) > pasture > wheat > chickpeas > fallow > wheat. However, on lighter sandier soils, the rotation would consist of pasture > pasture (green manure) > wheat > oats or simply pasture > pasture > wheat.

Additional benefits derived from cover cropping include weed and disease suppression. Approaches for managing weeds are discussed below; however, the main mechanism of weed suppression by cover crops appears to be resource competition, rather than factors such as allelopathy. The different occurrence of the fungal pathogen.

6

Pest and Pathogen Regulation in Natural versus Functional Cereal Grains

Since the 1980s, the proportion of cereal crops (wheat, barley, oats, rye) has steadily dropped in conventional agriculture in north-western Europe (except for France). The main reason is a reduction in price supports, resulting in such low prices that it is not profitable to grow cereals as intensively as was commonly done in this region (with straw shorteners, high fertilization, and frequent fungicide and insecticide applications). On organic farms, cereal crops are still profitable as they are grown less intensively. Moreover, they are essential crops in the longer rotation schemes. Some of these crops (oats, rye, and triticale) are also used as winter cover crops to reduce nitrate leaching. Because of the large differences in production practices for cereal crops at organic versus conventional farms in Europe, many comparative studies of farming systems have focused on these crops. Most comparisons have involved single (experimental) farms side by side. In a few cases, replicated experimental treatments were compared, and even fewer publications covered large surveys on many farms. From all these studies it is apparent that fungal root and foot rots caused by Fusarium species are similar or reduced in organic cereal crops. The same holds for take-all disease caused by Gaeumannomyces graminis and for eyespot and sharp eyespot caused by Pseudo Cercosporella hypotrochoids and Rhizoctonia cereals, respectively. In some studies, the lower disease levels at organic farms were associated with lower N application rates, higher microbial activity and diversity, or greater populations and/or diversity of soil microfauna. Another reason for suppression of root diseases in organic cereal crops could be enhanced competition by arbuscular mycorrhizae, which are generally more abundant in organic than in conventional crops, due to the lower available N and P in organically managed soils. The lower N application levels in organic cropping systems also have a tremendous influence on above-ground diseases and pests. Powdery mildew, snow mould and stripe rust of wheat were less severe in a long-term organic than in a neighboring conventional experimental farm in the Netherlands, despite regular fungicide applications in the conventional system. This was attributed to lower N levels in the organic wheat tissue, but could also be due to a more open canopy structure and less conducive microclimate. In an extensive survey of 150 Dutch wheat fields, most above-ground diseases (snow mould, powdery mildew, Septoria leaf and glume blotch, Fusarium scab) were less severe in the organic than in the conventional and integrated farming systems. The differences were significant for snow mould, glume blotch, and Fusarium scab. Contrary to the two- farm study mentioned above, there was no significant difference for stripe rust, while leaf rust and powdery mildew on the ear were significantly more severe in organic than in conventional farms. Incidences of diseases that were higher in conventional farms than in organic farms were again positively correlated with N application rates. The higher severity of leaf rust and powdery mildew on the ears of organic than on those of conventional wheat plants may also have been related to N concentrations, since N is released from soil organic matter in the, summer time, and can be higher in organic than in conventional farms at that time.

Plants high in N can also support large aphid populations and are often more susceptible to virus infection. Thus, when high soil nitrate concentrations coincide with aphid flights, the population may grow explosively. Unfortunately, organic farmers have little control over the time when N is released in soil, and in some seasons aphid populations may be as high in organic as in conventional farming systems. Leaf miners and cereal leaf beetles were generally more numerous in the conventional farming system and were associated with high N application rates. In an extensive field survey, no significant differences were found in populations of aphids, leaf miners and cereal leaf beetles between organic and conventional farms, but there were enormous variations among years. Nevertheless, it is important to try to keep mineral N concentrations minimal, also at organic farms. This can be done by focusing on organic matter build up over many years, and minimizing applications of additional organic fertilizers during crop growth.

The absence of fungicide use in organic cereal production has given rise to concerns about grain moulds and mycotoxins on organic cereals. These concerns are not always justified. Fusarium scab of wheat was less severe in organic than in conventional farms in the Netherlands. In several German studies, Fusarium contamination of grains and concentrations of deoxynivalenol (DON) were also lower in organic than in conventional farms, while in some French studies Fusarium head blight severity and mycotoxin levels were similar in organic and conventional wheat production. DON concentrations were also similar in organic and conventional grains in a British study, but were below the European threshold level. Average levels of ochratoxin A contamination (from Penicillium and Aspergillus) were similar in rye and barley grains from organic and conventional farms, but significantly lower in organic than in conventional wheat grains. Thus, there is no reason to believe that mycotoxins would constitute a problem in organic cereal products. The main reason for lower contamination levels at some organic farms may be the lower N contents and greater diversity of non-pathogenic fungi on the ears of unsprayed plants. Moreover, mycotoxin production per unit mycelium can be enhanced by the stress of certain fungicides.

Insect pests can often be kept in check by a greater diversity of non-herbivorous arthropods. An increase in beneficial insects during the conversion period on a biodynamic farm. A higher diversity of arthropods in organic cereal fields, accompanied by a lower arthropod abundance compared to conventional cereal fields. In many other studies, a relatively high arthropod abundance in organic compared to conventional wheat fields was associated with a higher arthropod diversity in organic wheat fields. Measured similar levels of pest butterflies in organic versus conventional farmland, but found significantly more non-pest butterflies in organic farmland. Total numbers of epigeic predatory arthropods were also highest in organic farming systems compared to pesticide intensive agricultural production systems. Among epigeic predators, carabid abundance and species richness were higher in organic cereal fields, while staphylinid and spider abundance and species richness were generally similar in different management systems. The reasons for the frequently greater arthropod diversity in organic farms can be found in the greater variety of food sources associated with greater plant diversity within fields and in surrounding habitats.

Greater plant diversity in the field can also have benefits for disease control in cereal crops. Different cereal crops, for example barley and wheat, have sometimes been grown in mixtures resulting in a reduction in barley powdery mildew by increasing the distance between susceptible plants. However, mixtures of wheat and field beans resulted in higher powdery mildew severity on the wheat crop as the bean density increased; this was probably as a result of the greater N content of the wheat plants. Cultivar mixtures with different resistance genes have also been very effective in controlling different barley powdery mildew pathotypes. The use of cultivar mixtures was a very effective and widespread practice in the former Eastern Germany and is now practiced on a large scale in rice production in China.

Unfortunately, organic growers have not yet widely adopted this practice. Another form of mixed cropping, relay cropping of clover between cereal plant rows, is gaining popularity in Europe. This practice helps control weeds, supplies N to the next crop, and contributes to disease control, particularly of take all caused by Gaeumannomyces graminis.

Thus, organic practices that promote biodiversity both above ground and below ground tend to enhance the resilience of the arable farming system so that many pests and diseases are controlled by natural enemies. These practices include an extensive rotation, soil organic matter management fostering high turnover rates and relatively low residual mineral nutrients with minimal fluctuations, crop or cultivar mixtures, and varied field margins. Although this is a desirable scenario for cereal production, it is not always practiced, and the great variation in organic farming practices understandably results in large variations in pest and disease intensities.

Soil Fertility in Organic Farming Systems

All farming systems depend on the maintenance of soil fertility for environmental and economic sustainability. The effective management of soil fertility is based on understanding processes associated with its chemical, physical and biological components, but greater emphasis has traditionally been given to soil chemical fertility. Soils vary greatly in their inherent physical and chemical characteristics and thus in their capacity to support biological activity. Of course, agricultural practices also affect soil fertility but the magnitude of change depends on specific soil and environmental conditions. Therefore, local knowledge is necessary to interpret the effects of farming practices on soil fertility.

Although organic farming has features of 'sustainable' agriculture, it cannot be assumed that organic management per se is more or less sustainable than some other farming systems. While all organic farming systems operate without the use of synthetic chemical inputs, many other practices used in organic farming are employed on both 'organic' and 'conventional' farms. Many conventional farming systems effectively address issues related to sustainability such as nutrient use efficiency, appropriateness of rotations to landform and soil type, conservation tillage, biodiversity and use of legumes. Thus, there is a continuum in the level and nature of changes in soil characteristics associated with agricultural management practices, rather than two distinct categories of 'organic' and 'non-organic'. Consequently, the concepts of sustainability and soil fertility management in organic farming systems have parallels in systems that use synthetic chemical inputs.

A key feature of organic farming systems is that they are, by necessity, very dependent on soil biological fertility, which in turn influences aspects of chemical and physical fertility. Organic farming systems aim to create temporal and spatial diversity in plant, animal and microbial life. Particular emphasis is placed on creating a diverse microbial community to maximize nutrient use efficiency from poorly soluble fertilizers and promote beneficial soil physical processes. In contrast, those farming systems that primarily use synthetic chemical fertilizers to enhance soil chemical fertility may bypass important aspects of soil biological fertility that could contribute to the chemical, as well as to physical fertility of soil. Current standards for organic practice may best suit chemically fertile soils in temperate climates rather than more impoverished soils in other regions. Globally, poor soils are generally dominant in tropical latitudes and the most chemically fertile soils primarily occur in temperate areas. Although differences in distribution of basic plant nutrients influence the extent to which farming systems are dependent on inputs, local exceptions occur within these global generalizations. Certified organic farming practices evolved in regions with chemically fertile soils (especially in Europe and North America). In contrast, precursors of modern organic farming practices have been successful on chemically infertile soils for hundreds of years (e.g., in tropical soils in Africa, South America and Asia) and may include high inputs of organic matter (including animal and human waste). At the other end of the scale, great expanses of arable land are certified 'organic' primarily because of the impracticality of using synthetic inputs (e.g., vast areas of arid Australia). On a global scale, inherent soil chemical infertility is associated with the greatest plant species diversity. The results of field trials and paired comparisons (e.g. 'organic' versus 'conventional' farms) evaluating effects of organic farming systems on components of soil fertility need to be interpreted with caution. Long-term studies provide an opportunity for soil biological processes to stabilize. However, few of the published field trials evaluating the effect of farming systems on soil fertility have been run for longer than eight years and most have been conducted in regions with naturally chemically fertile soils and temperate climates. Also, the evaluation of soil fertility in organic farming systems needs to be made at the level of specific management practices but this is not always practical. Both direct and indirect effects of specific management practices on the components of soil fertility should be considered to allow more complete interpretation of the impacts of soil disturbance resulting from changes in agricultural practices on soil fertility. In this review, we consider components of soil fertility in the context of organic farming systems and how certified organic practices are used to manage soil fertility. Data from long- term field studies evaluating organic farming are discussed in terms of the components of soil fertility. The concept of soil fertility is also viewed in relation to its relevance to the application of organic farming practices across landscapes and soil types in different climatic zones.

Components of soil fertility the use of the term 'soil fertility' is of little value unless the physical, chemical and biological properties of soil that contribute to its fertility are all considered. There is widespread agreement as to general standards for 'good' soil physical and soil chemical fertility (which take into account different soil types), but standards for 'good' soil biological fertility are more difficult to define. Furthermore, in contrast to chemical fertility, there are few commercial services available for farmers to measure soil physical and biological properties, and so the use of this information is not wide-spread. The greater emphasis on biological processes in organic farming means that the chemical fertility of the soil depends to a large degree on how management practices facilitate beneficial biological processes.

Soil Chemical Fertility

The chemical fertility of soil reflects its capacity to provide a suitable chemical and nutritional environment to plants and to support biological and physical processes. The maintenance of soil chemical fertility in organic systems depends strongly on processes that govern transformations from fixed to soluble forms of nutrients such as mineralization of organic matter and dissolution of minerals. Although these processes also occur in conventional agricultural systems, organic farms rely on them to a greater extent. Traditional methods for predicting fertilizer applications for building and maintaining soil chemical fertility may not be appropriate in organic farming systems. Most tests of chemical fertility were developed to predict nutrient release from highly soluble sources using relationships developed between chemical extracts of soil nutrients and plant uptake. These relationships may not be directly applicable to organic farming systems where:

- Nutrient sources are predominantly organic or poorly soluble and therefore slow to become available;
- Nutrient availability is more dependent on dynamic soil biological processes and
- Release and uptake of nutrients occurs without demonstrable changes in soil chemistry because nutrients are rapidly taken up by plants or soil microorganisms without accumulating in the soil solution.

For example, found that in organic farms, although soil inorganic N was one-quarter that of conventional farms, organic N mineralization potential was three times larger. Similarly, concluded that routine soil tests would not have been sufficient to predict phosphorus (P) fertility due to increased mobility of P ions in a bio-dynamic system. Options for monitoring soil fertility in organic farming systems may include ongoing (but time-consuming) quantification of trends in soil and plant analyses. Also, sequential tests may prove more useful for predicting the availability of nutrients to plants through a growing season in organic farming systems. The Organic Advisory Service in the United Kingdom performs a sequential P extraction to determine soil reserves, plant available and water-soluble P.

Combining soil analysis with nutrient budgets provides another way to monitor soil chemical fertility in organic farming systems, particularly in view of the concern that organic farming systems are mining soil nutrient reserves. Nutrient budgets quantify nutrient inputs and outputs from a defined area in a given time. Surpluses indicate the potential for nutrient losses to the environment and deficits suggest soil reserves of nutrients are being exploited and question the sustainability of the system in terms of soil chemical fertility. Nutrient budgets may also be used to determine nutrient use efficiency. However, nutrient budgets may be inaccurate because it is difficult (i) to estimate the nutrient contents and application rates for manures and composts and (ii) to directly measure N fixation and leaching on commercial farms. Nitrogen fixation is generally estimated from published studies, and N leaching is often excluded from nutrient budgets.

Soil Biological Fertility

Soil biological fertility refers to soil processes involving organisms that improve plant growth both directly (e.g., symbioses with root nodule bacteria and mycorrhizal fungi) and indirectly, through their effects on soil chemical fertility (e.g., organic matter mineralization and mineral dissolution) and physical fertility (e.g., soil aggregation). Soil biological fertility can be quantified by measuring the size, activity, diversity and function of communities. However, there are no agreed standards and the full implications of increased populations, activity and diversity of soil organisms on soil function and plant growths are not known. Measures of soil biological fertility may be useful as indicators of long-term changes in overall fertility because they respond rapidly to changes in soil conditions. Learning how to manage beneficial soil biological processes may be a key step towards developing sustainable agricultural systems. Central to organic farming is the aim of optimizing plant production by maintaining a rich biological diversity in the soil. A soil's capacity to support biological fertility is determined by inherent physical and chemical characteristics as well as management practices. The Sustainable Agriculture Farming Systems Project (SAFS) trial showed that the relative importance of several environmental variables governing the composition of microbial communities were ranked in the order: soil type > measurement time > specific management practice > management system > spatial variation. Management practices that may be used to maximize the benefits of soil organisms include organic matter additions, increasing plant diversity, reduced tillage practices and certain soil amendments. Without increased inputs of organic matter, neither organic nor conventional farming systems can expect to increase the size or activity of the soil biological community.

Debate surrounds the potential role of soil biological processes in maintaining soil chemical fertility in organic farming systems. In theory, the chemical fertility and sustainability of organic farming relies on soil biological fertility to a greater extent than in conventional farming systems. The fertilizers permitted in organic farming systems are relatively insoluble in soil (IFOAM 2002) and the processes that govern their capacity to release nutrients are mediated or improved by soil organisms. Organic farming may alter the function of the soil microbial community, increasing its ability to release nutrients from organic and poorly soluble sources, thereby compensating for the absence of soluble nutrient inputs. This may be related to a 'priming effect'. However, found that despite differences in microbial community com- position and functional diversity, there was little difference in gross mineralization rates between the organic and conventional arable farming systems. Other field work showed that there was no difference in the response of crops grown in organically and conventionally managed soil to rock phosphate application, indicating that organic management had not increased the availability of rock phosphate to crops. Further research is needed to scientifically investigate the exact nature of biological activities under long-term organic and conventionally managed farming systems, especially in soils that are highly weathered.

Practices that enhance function of arbuscular mycorrhizal (AM) fungi and root nodule bacteria have potential to improve some aspects of soil biological and chemical fertility. Plant breeding for greater dependency on these symbioses could be equally important. AM fungi have the potential to access P in soil pores unavailable to plant roots, but it is not known whether this mechanism has a significant effect in overcoming P deficiency. AM fungi can also improve plant availability of poorly soluble P sources such as rock phosphate and P adsorbed in soil by enabling plants to have access to P as soon as it is released into the soil solution. Under organic farming practices, the extent of root colonization by AM fungi may be greater due to decreased use of soluble P fertilizers and the diversity of AM fungi can be greater. However, the benefits of AM fungi to plant production and sustainability are very difficult to measure and appear to be environmentally dependent. It may be possible to selectively increase the abundance and activity of certain soil organisms using agricultural inputs. For example, simple organic compounds such as sugars and complex humid substances may stimulate microbial activity, leading to short term increases in biological activity and, potentially, to nutrient release and improved physical fertility. Demonstrated that specific fungi and bacteria occurred preferentially on different silicate minerals. This observation is very significant if microorganisms preferentially stimulated in this manner have beneficial effects on soil fertility. While further understanding of the dynamics and diversity of soil biological processes is necessary in all farming systems, a soil may have 'too much' biological activity. This can occur if organic matter is continuously disturbed and exposed to rapid degradation, with subsequent loss of its value as a slow-release nutrient source and contributor to maintaining or improving soil structure.

Managing Soil Fertility in Organic Farming Systems

Underlying Principles

The practices used to manage soil fertility in organic farming systems should be understood in terms of the aims and underlying principles on which they are based. This has been discussed in various reviews of aspects of organic farming systems and in organic production standards. To achieve the long-term goal of sustainability, organic farming systems aim, as far as possible, to be self-sufficient for nutrients and organic matter by producing and reusing materials on-farm. Therefore, practices that facilitate the efficient re-use of nutrients and organic matters within the farm are stressed. Also, organic production standards only permit non-synthetic fertilizers that are poorly soluble in the soil solution. Nutrient management in organic farming systems is not simply replacement of soluble fertilizers with insoluble fertilizers. To build and maintain adequate soil fertility, organic farming systems must integrate management practices such as those discussed below.

The unscientific nature of the ban on manufactured inputs in certified organic farming systems has been criticized, but the contrary practice of seeking to maximize crop yield using synthetic inputs regardless of the environmental consequences is not scientifically based either. The design of treatments in shows how a farmer's bias towards achieving certain objectives from production influences management choices. However, it is not the purpose of this review to justify the underlying assumptions of organic farming systems, but rather to discuss their effects on soil fertility.

Although it has been claimed that organic farming systems are fundamentally different from other systems, this cannot be substantiated from a soil fertility perspective. The management of soil fertility in organic farming systems is variously characterized by mixed livestock arable systems, crop rotations, legumes, organic matter inputs and the use of fertilizers that are not readily soluble in soil; similar practices also characterize conventional farming systems that address sustainability issues, including the use of legumes to manage plant and animal nutrition in broad acre livestock–crop production and crop and forage rotations to manage pests and soil fertility.

Livestock

Livestock can directly improve soil chemical and biological fertility by introducing organic matter and nutrients in manure and urine. Although animal traffic may decrease some aspects of soil physical fertility, the reduced tillage, dense rooting, increased root exudation and soil organic matter content associated with pasture phases increases soil aggregation and biological fertility and reduces erosion. Livestock can also help to control weeds which may reduce chemical fertility by competing with plants for nutrients. The use of manure helps to achieve the aim of self-sufficiency in nutrients and organic matter, but it is not always available in large quantities and its management during storage and handling needs to minimize gaseous and leaching losses of nutrients. Rotations that include livestock can be a key component of sustainable farming systems. The retention of mixed farming in organic farming systems and the associated increases in spatial and temporal habitat heterogeneity was partly responsible for increased diversity of organisms observed on organic farms, from soil microbes to mammals and birds. Without livestock, it can be difficult to manage nutrients, particularly N, in organic farms but dairy farms may be an exception to this because of their susceptibility to N losses.

Legumes

Legumes are a fundamental component of organic farming systems (in pastures, green manures, cover crops or food crops) because they reduce or eliminate the need for external Fertilizers providing, they are effectively modulated. The sustainability of legume uses to supply crops with N in either organic or conventional farming systems depends on the:

- Fixation of sufficient N in the legume biomass;
- Ability of the soil community to increase mineralization of organic N; and
- Capacity of farming practices to maximize the beneficial soil fertility and environmental effects of legumes and minimize their negative effects (e.g., increased acidity and N leaching).

Nitrogen fixation by legume biomass may be efficient but the availability of the residues to subsequent crops could be improved. Farm-scale nutrient budgets for 88 organic farms in temperate climates and found that although all of the N budgets showed a surplus, N use efficiency was low (average 0.3). Concluded that although legumes had the potential to supply sufficient N to crops, there was usually a shortfall. Where legume pastures are incorporated only once every few years, organically produced crop residues and manures tend to have low N contents and slow mineralization rates. Improving legume-N availability to subsequent crops could be achieved by (i) careful timing of organic residue incorporation, (ii) management practices that improve the quality of legumes as microbial substrates and (iii) matching crop type with N mineralization dynamics.

For legumes to maintain adequate N availability to plants, reductions in soluble N inputs must be compensated for by increased mineralization of organic N. Potentially mineral sable N is frequently higher in organic farming systems. Attributed the higher average rates of N mineralization (30%) for integrated compared with conventional farming to higher soil organic matter contents, but there was no difference in gross N transformations between a pair of organic and conventional arable farms. Mite diversity in a conventional soil under wheat was correlated with gross N immobilization, so this may also be relevant in organic farming systems because mites can increase the diversity of soil fauna.

The sustainability of using legumes to supply crop requirements for N also depends on the capacity of organic management practices to maximize beneficial effects of legumes while minimizing the potential for N leaching. Intercropping with legumes can increase the efficiency with which soil nutrients are used. Increased use of legumes in farming systems because of beneficial environmental effects including improved soil structure, erosion protection, increased biological diversity, and stimulation of rhizosphere organisms, acidification of alkaline soils and reduced energy use and carbon dioxide (CO_2) production on and off the farm. However, asynchrony in plant demand for N and its release from organic matter can cause nitrate leaching, especially post-harvest, and the acidifying effect of legumes is detrimental in acid soils nitrate leaching from organic and conventional farms. They concluded that although the average nitrate leaching over a rotation was lower in the organic systems investigated, when the lower intensity of N input in organic farming was taken into account, there was no difference. In this evaluation, there were insufficient data to compare nitrate leaching based on yield. Management options for controlling N leaching were well-managed organic systems have the potential to use excess N though practices such as catch crops and by intercropping.

Rotations

Rotations facilitate processes that alleviate some of the fertility constraints to production that are addressed in conventional farming systems by use of synthetic inputs. Organic rotations may include greater use of green manures and cover crops, emphasize different regional crops, and have longer pasture phases than on conventional farms, leading to higher plant diversity in space and time. Rotations are also used to manage soil physical fertility by emphasizing inputs of organic matter from pasture phases, green manures or cover crops.

Organic matter Management

Organic matter incorporation into soil in organic farming systems is pivotal to increasing chemical fertility and improving soil structure and this has been extensively investigated. Organic farming systems emphasize frequent additions of diverse sources of organic matter from catch crops, crop residues, manures, some forms of organic fertilizer and perennial crops. Numerous paired comparisons of organic and conventional farms demonstrated that under organic farming practices, soil organic carbon (C) content increased by up to 30%. Examples of more substantial increases in soil organic C content (up to 200%) are probably due to the comparisons being limited to a single organic and conventional farm or to assessment immediately following large inputs of organic matter. Apart from benefits to chemical and physical fertility, whether or not organic farming systems increase organic matter inputs relative to conventional farming systems determines their ability to increase soil biological activity. Increases in the abundance, activity and diversity of soil organisms under organic management are primarily caused by increases in the amount and quality of organic matter inputs. In some situations, the higher soil organic matter contents under organic farming increased microbial biomass and activity by increasing soil water content.

The difficulty in increasing soil organic matter content in some environments may threaten the sustainability of organic farming systems. In semi-arid environments with high temperatures and low precipitation, C and N inputs to soil may be low and organic matter content unrelated to soil texture. Also, the degradation of organic inputs may be very rapid in these environments if the soils are sandy, because they offer little protection to organic matter. Several Australian studies showed that compared to conventional farms, organic management did not increase soil organic C in dry land grain-livestock production in southern Australia. In contrast, soil organic C did increase in irrigated organic farming systems. The reasons organic farming may not always increase soil organic C contents include:

- The larger yields and more intense animal production in conventional systems increasing organic matter inputs;
- The smaller C:N ratio of organic leys causing them to decompose more quickly; and
- More intensive cultivation for weed control.

Organic farming systems that fail to increase levels of soil organic matter cannot be expected to increase overall soil biological activity. Further research is needed to investigate the role of organic matter and soil biological fertility in the sustainability of organic farming systems in semi-arid environments with highly weathered soils, especially in relation to organic certification standards.

Fertilizers

The fertilizers permitted in certified organic farming systems, and in some cases the amount that may be applied, are restricted by organic certification standards. They are loosely divided into two categories: (i) naturally occurring geological resources (minerals) and (ii) organic materials. Minerals permitted as fertilizers in organic farming systems include lime, gypsum, rock phosphate, guano, elemental sulfur (S), dolomite and various ground silicate minerals. Organic materials include those produced on farms such as green manure, animal manure and compost, as well as off-farm sources such as fish, blood and bone meal, seaweed extracts and microbial products. Complete lists of fertilizers permitted in organic farming systems can be found in the various organic certification standards. Some restrictions may not be based on scientific evaluations. For example, the allowable level of organic matter (e.g., as compost) may be less than that required in poor sandy soils. In Australia, the National Association for Sustainable Agriculture Australia (NASAA) limits off-farm manure and compost applications to 15 and 20 t ha^{-1} y^{-1} respectively (NASAA 2003) and the European Union standards limit manure application to 170 kg N ha^{-1} y^{-1}.

Estimation of the relative effectiveness of nutrient sources can be a useful way in which to estimate their efficacy as fertilizers. The relative effectiveness of alternative nutrient sources is usually calculated by comparing the yield plateau of the response curve of the fertilizer in question to a soluble source of the same nutrients. For minerals used as nutrient inputs in organic farming systems their relative effectiveness is almost always <1 due to low solubility in soil. Organic matter inputs can also be evaluated in terms of their relative effectiveness based on their recalcitrance, but of equal importance is the extent to which they are physically protected from degradation in soil aggregates, which would be different in different soil types. Silicate minerals and rocks composed of silicate minerals are used in organic farming as fertilizers and soil ameliorants. Silicate minerals are the main components of igneous and many metamorphic rocks and vary in their composition and dissolution rates. Dissolution is favored by minerals of small grain size and large surface area and is greater in soils with low pH, high moisture and temperature and soil solutions that are not in equilibrium with mineral surfaces. Silicate minerals are most suitable as fertilizers in highly weathered, tropical soils where acidic, nutrient deficient soils and heavy rainfall events favor dissolution and also in leaching soils, particularly sands. In addition to a soil's capacity to dissolve silicate minerals due to inherent chemical and physical characteristics, plants and microorganisms increase silicate mineral dissolution. They release organic ligands which attack mineral surfaces and form complexes and lower soil pH by releasing H^+ ions and organic acids into the soil. Nutrient uptake by plants prevents equilibrium between minerals and soil solution from being reached, which stimulates further dissolution.

Research on plant uptake of nutrients in silicate minerals has focused on release of K and several researchers have concluded that silicate minerals have potential as slow release fertilizers, but rocks and minerals high in silica (e.g. granite and feldspar) may be poor sources of potassium. When the dissolution of ground granite in 20 acid soils from Western Australia was measured, few soils showed an increase in exchangeable calcium (Ca) or magnesium (Mg) and nine soils showed an increase in exchangeable potassium (K). Incubation experiments and field and pot trials have shown that between 1 and 10% of the K in feldspar is released up to 14 months after application and the relative effectiveness of granite as a potassium fertilizer was <0.14 compared to KCl.

Silicate minerals may also provide plants with Ca, Mg and some micronutrients. Studies have measured the release of Ca and Mg from silicate minerals and rocks such as amphibolite, basalt, diabase, dunite, gneiss, granite, phenolite, serpentine, syenite and a volcanic ash that on an equal weight basis the Mg content of tobacco, white clover and ryegrass plants increased in the order dolomite < ser- pentine < dunite. Application rates between 1 and 50 t ha^{-1} of basalt increased the exchangeable K, Ca and Mg in seven highly weathered tropical soils in Queensland, Australia. For five soils, 5 t ha^{-1} of basalt increased exchangeable cations but on one soil, 1 t ha^{-1} of basalt was sufficient to increase exchangeable Mg. Silicate minerals also contain plant micronutrients but very little work has been done to assess their plant availability. Certain silicate rocks in both water and 0.1 M hydrochloric acid (HCl) released iron and manganese to a greater extent than copper and zinc. In contrast to their use as nutrients, silicate minerals have been advocated as soil ameliorants especially for lateritic soils in tropical climates. Silicate minerals increase soil pH, although not as effectively as lime and to varying degrees on different soil found that after nine months, granite applied to a highly weathered soil at 300 t ha^{-1} increased the soil cation exchange capacity from 9 to 14 meq/100 g of soil. Ground silicate minerals may also increase water-holding capacity. The highly weathered soils in tropical climates have most to gain from the use of silicate minerals as soil ameliorants.

For organic farming systems that do not have access to adequate quantities of manure to balance P lost from soil in harvested products, rock phosphate (and guano) may be used. The use of phosphate rocks and factors affecting their relative effectiveness were. Their relative effectiveness is affected by mineral properties (reactivity, particle size, surface area), soil factors (pH, titratable acidity, P and Ca availability and retention, sand content, organic matter content, moisture and temperature) and plant factors (P and Ca demand, root structure, rhizosphere pH) . Although many of the factors affecting rock phosphate dissolution are known, it is not easy to predict their relative effectiveness. Sources of rock phosphate have unique reactivity, for which citrate solubility has become a standard test. In an evaluation of the relative effectiveness of 14 rock phosphates, differences in rock phosphate reactivity caused a ten-fold difference in dry matter yield between the least and most reactive. However, the relative effectiveness also depends on the particular soil and plant factors and can result in a rock phosphate being as effective as superphosphate or nearly inert.

There is potential to increase the relative effectiveness of poorly soluble minerals by managing soil biological processes. As already mentioned, plant availability of rock phosphate can be improved by AM fungi, and some soil micro-organism can solubilize P and increase the dissolution of silicate minerals (Barker *et al.* 1997). showed that in a P limiting environment, micro-organisms selectively colonized the surface of minerals containing P. The ability of these processes to increase nutrient availability to plants might be maximized if specific practices and inputs can be found to target and increase populations of soil organisms involved. Alternatively, plant species and varieties may be chosen that cause the greatest mineral dissolution. Also, because the mechanisms by which plants cause mineral dissolution can be stimulated by nutrient deficiencies, breeding plant varieties in suboptimal nutrient conditions may produce varieties with greater capacity to dissolve minerals and to form associations with more beneficial AM fungi.

Other procedures for increasing the effectiveness of silicate minerals and rock phosphates involve altering the minerals themselves. High-energy milling (ball-milling at high energy intensities) causes changes to mineral structure and bonding, and is acceptable under organic production standards. Compared to unfilled rock phosphates, high-energy milling increased the relative effectiveness of five rock phosphates by up to three times. High energy milling for 120 minutes increased the release of Ca and Mg from basalt and dolerite from 2% to about 18%. The rapidly dissolved K from feldspar was increased from 0% to 27% by 120 minutes of high-energy milling. Co-composting rock phosphates and silicate minerals may also increase their relative effectiveness, by increasing their exposure to microbial processes that cause their dissolution, but more rigorous research is needed to confirm this. Others have demonstrated that shaking silicate minerals in organic extracts that are readily available to farmers, such as brewer's yeast and cattle urine, increased the amount of Ca, Mg and K released from silicate minerals.

Very little work has been published about micronutrients in organic farming systems. concluded that organic farming might be unsustainable on soils deficient in micronutrients unless measures are taken to supply them in fertilizers. Seaweed extracts can provide plants with micronutrients and opportunities exist to optimize their use. However, little is known about the optimum time to harvest seaweeds or the nutrient content of different species and their nutrient content is not high enough to meet crop demands. Furthermore, they may stimulate plant growth and yield through action of plant hormones rather than nutrients. Micronutrient nutrition of livestock can be maintained by growing pasture species rich in trace elements, but this requires further investigation. Morganic farming has the potential to exploit soil reserves of nutrients and may, therefore, be considered unsustainable. In some cases, nutrients 'mined' in organic farming systems were previously added as fertilizer when the land was farmed using conventional practices. Published nutrient budgets comparing organic and conventional farms have focused on N, P and K. They indicate that it is possible to balance nutrient budgets in organic farming systems, but that budgets are often negative because nutrients are not adequately replaced. Outside of temperate climates, it may be more difficult to maintain levels of soil nutrients because manure is less readily available. Several Australian studies have reported that P availability limits production on organic livestock–crop farms and threatens the sustainability of organic farming systems in Australia several of the organic grain–livestock farms in their study did not use fertilizers and that those that did used only small amounts of poultry manure (40 kg ha^{-1}). In a comparison of organic and conventional mixed livestock–cereal farms in New Zealand, the N budgets were balanced by the use of legumes, while P and S budgets were balanced on the farms that used compost, rock phosphate and elemental S and negative on the farms that did not. Clearly, this is an area that requires further research and highlights the potentially vital role of soil microorganisms in organic farming systems if they can be managed to more effectively cycle nutrients.

Considerable commercial attention is being given to development of microbial products that can be added to soil to stimulate release of nutrients and benefit plant growth in organic farming systems. Many other forms of organic (plant, animal and microbial) fertilizers are permitted, but most have not been scientifically investigated under field conditions. Further research is required to validate the capacity of these materials to enhance soil fertility in the long term and in combination with other management practices, using well-replicated scientific studies, otherwise claimed effects will remain anecdotal.

7

Experimental Studies in Organic Farming

Like all farmers, organic farmers look for the best varieties for their farm conditions. The pool of varieties to choose from consists, in many cases, of conventional varieties, bred for conventional agricultural systems in which high levels of artificial fertilizers and agrochemicals are widely used to control or overrule external factors. For such farming systems, conventional plant breeding can aim at optimizing yield at high input. As organic farmers refrain from such chemical inputs, it seems logical that requirements for varieties suitable for organic farming systems should be based on optimizing yield at lower input and stabilizing yields under less controllable conditions. Most variety trials, though, show that the organic sector can profit from the progress in plant breeding and that modern varieties also fulfill some of the important needs of organic farmers. Not only has the yield potential of varieties increased but also the level of disease resistances has improved in some crops, such as tomato.

That many organic farmers in developed countries use modern varieties does not imply that those varieties are the best ones for Organic Farming.

Given that attention is often focused on developing other agronomic areas of their farming systems, organic farmers have long accepted their dependence on conventional breeding. However, now more attention is being paid to the question of how to develop better-adapted varieties for organic farming systems. This question is even more urgent since genetic engineering has become important in conventional plant breeding and being dependent on conventional breeding would therefore no longer be an option. The organic sector realized that it is not only concerned about the variety traits as such but also about how varieties are bred and propagated, and thus whether breeding and propagation methods comply with both the ecological and ethical principles in Organic Farming.

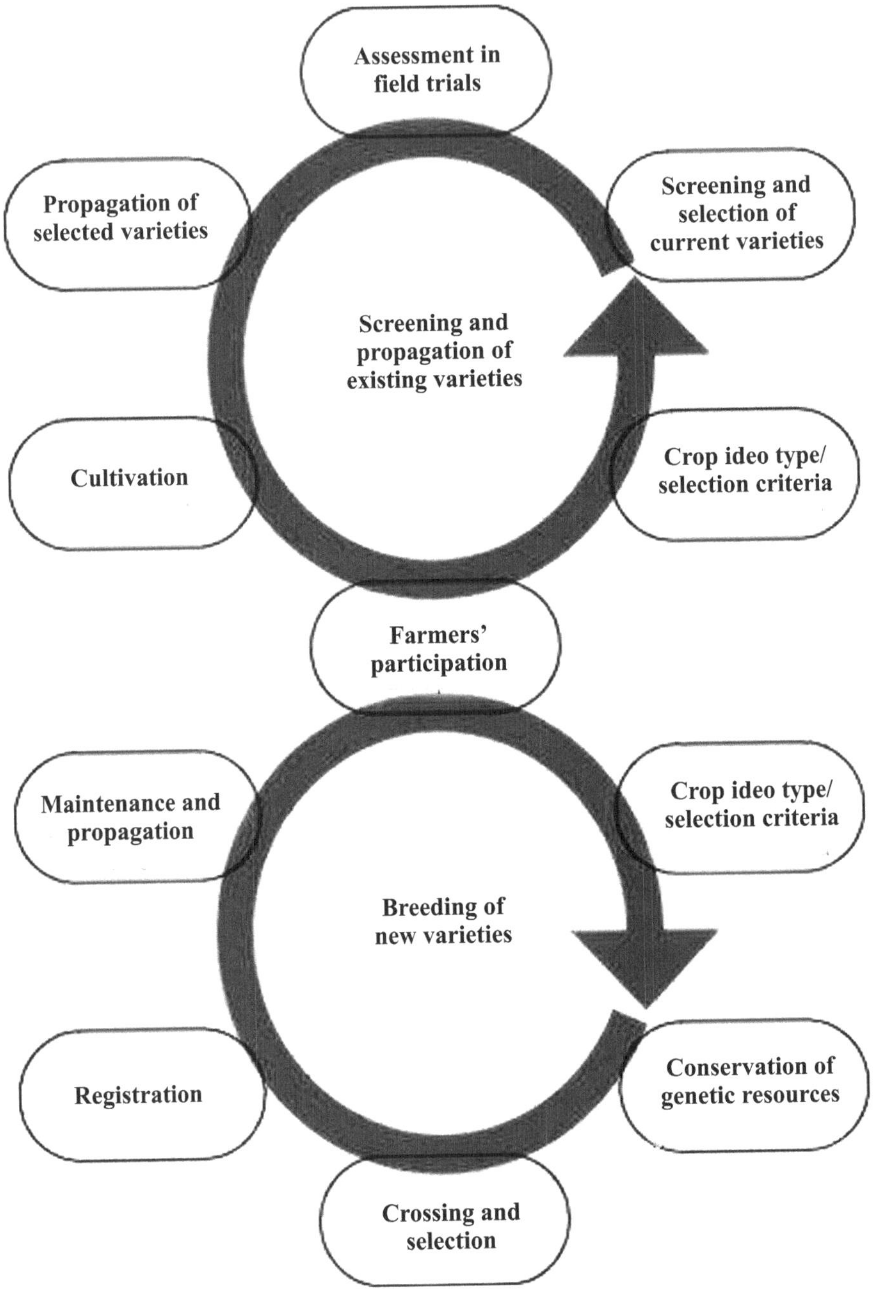

Fig. 7.1: A flow chart of organic plant breeding and propagation

Such principles applied in the organic sector can be reflected in the concept of naturalness, which encompasses three approaches:

1. The non-chemical approach;
2. The agro ecological approach; and
3. The ethical approach in which the integrity of life is taken into account.
4. To arrive at varieties and seeds produced in accordance with the needs and principles of Organic Farming the whole plant breeding and propagation chain will need additional research efforts and practical steps (Figure 1.3). It involves both defining crop ideotypes, and screening and selection of existing varieties for propagation, and in situ conservation of genetic resources, developing crop ideotypes and selection criteria for breeding programs, including registration and maintenance activities.
5. As the development of each new variety takes at least ten years and a lot of resources, the development towards varieties adapted to organic farming conditions will have to be taken in steps (Table 7.1).

Table 7.1: A time schedule for steps and results towards organic varieties and seed production

Time	Approach	Activity	Product
Current (2005)	(Primarily) non-chemical approach	Defining desired traits Variety screening Refrain from postharvest, chemical seed treatments No use of transgenic engineering	Untreated seeds from conventional varieties
Short-term (2005-10)	Non-chemical approach Including an agroecological approach	Organic propagation Organic seed treatments	Conventional varieties, but organically propagated (organic seeds)
Mid-term (2010-20)	Non-chemical approach Agroecological approach	Organic VCU-testing protocols Conventional plant breeding programs including some low-input selection criteria	Low-input varieties Organically propagated (organic seeds)
Long-term (beyond 2020)	Non-chemical approach Agroecological approach Including the concept of integrity of life	Breeding/selection and maintenance under organic farm conditions With techniques that allow crossing, pollination, fertilisation and seed formation on the whole plant itself	Organic varieties Organic seeds

The first step in the short run is to define desired characteristics and select the best performing varieties from the existing assortment. The second step in closing the organic production chain is stimulating organic seed production of the above selected varieties. This is forced in Europe by the EU regulation 2092/91 that allows no more derogation (i.e., formal substitution of an organic input with a conventional input where no organic input can be found) for the use of conventionally propagated seed from 2004 onwards. This counts for those crops with already sufficient number of varieties appropriate for organic farming. Similar regulations have been enforced in the United States of America (USA) as part of the National Organic Program (NOP). Another important step is to influence the national procedures and protocols for testing of varieties for value for cultivation and use (VCU) and the release of new varieties, which is important to enlarge the chance of varieties entering the market with traits that are important for the organic sector. The step for the long run is developing breeding programs for the improvement of varieties adapted to the requirements of the organic food production.

For rapid and balanced progress in all these steps it is important to involve all stakeholders (farmers, traders, breeders, policy makers) in a participatory way in (research) projects to develop appropriate concepts and strategies and to solve practical obstacles. In this chapter we specify the need for better adapted varieties, and describe possible ecological and ethical approaches to develop such varieties, and the attempts made to develop organic propagation of varieties to close the chain and refrain from conventionally produced inputs.

Variety Characteristics

Variety choice in organic farming is an essential factor of successful production. But there is a significant lack of information on the performance of varieties under organic conditions. As conventional plant breeding aims at optimizing yields under high inputs, the conventional variety testing selects for high performance under high input conditions. As the diversity of organic farming systems and conditions is large and results in a larger genotype– environment–management (G.E.M) interaction than in conventional agriculture, the organic farmer is more interested in varieties that have the ability to adapt to variable conditions and perform well with stable yields in different years at the specific site. In many countries there are variety trials under organic conditions but not for every crop and every year. Some are conducted by farmer groups, others by research stations or by commercial seed companies.

Yield Stability and Plant Health

Long-term trials show that yields fluctuate much more in Organic Farming than in conventional agriculture. This is reflected by a larger coefficient of variation. For some crops this year-to-year variation has become one of the most important factors that limit the growth of the organic market share. To reduce risks and increase yields the organic farmer first of all aims at system stability. That can be done through improving agronomic strategies and selecting appropriate varieties. For many crops one of the limiting factors influencing yield stability is the suitability of the available varieties. Organic farmers would rather have a variety with a reasonable yield and good stress tolerance than a variety that promises higher yield but largely loses that yield increase because of, for instance, disease susceptibility. Reliable varieties are those that can cope with unfavorable weather and soil conditions (e.g. fluctuating mineralization rate of nutrients, assessed over several years). Farmers look for varieties that are, in general, more flexible and robust to support the self-regulating capacity of the organic farming system.

Conventional breeders often think that organic farmers require an absolute disease resistance, but in many cases tolerance or partial resistance can be sufficient. Organic farmers are used to searching for ways to reduce risks by an array of solutions at different levels (farm, crop, variety) and also adopt such a strategy in their search for appropriate combinations of supporting variety characteristics. There are different variety characteristics that support yield and yield stability through optimal plant health (Table 7.2). Plant health can be defined as a result of the following components:

1. Escape – the crop is not exposed to a certain pest or disease;
2. Resistance – the crop is exposed to the pest or disease but is not affected;
3. Tolerance – the crop is affected by the pest or disease, but with a limited effect on the yield; and
4. Ability to recover – the affected crop is able to recover to a certain extent.

Product Quality

In the organic sector, the non-chemical approach also has effects on traits concerning product quality, such as long-term storability without chemical sprouting inhibitors for onions and potato. In the case of fruit, a skin with a sufficiently thick natural wax layer, instead of a skin coated with synthetics, can positively influence longer storability.

Table 7.2: Variety characteristics supporting plant health in organic crop production

Mechanism	Variety characteristics
Escape	Shorter crop growth period and/or early ripening to be able to avoid the critical infestation period, or to have enough yield before the infestation becomes too severe (potato/lateblight; onion/ downy mildew; carrot/carrot fly).
Resistance	One or more monofactorial and multifactorial, durable resistance traits against pest and disease affecting yield and/or quality (scab/ apple, late blight/potato, lettuce/downy mildew, yellow and brown leafrust/wheat); weed competition by an allelochemical ability.
Tolerance	Morphological or physiological traits to support reducing the risk of a rapid expansion of the infestation, such as for cereals: long stem, ear high above flag leaf, ear not too compact; leaf erectness/ onion; hairy and tougher leaves against aphids/potato; wax layer on leaves against fungi/onion/ cabbage. Stress tolerance in general by good adaptation to low(er)-input and organic inputs, ability to cope with fluctuating N- dynamics, efficient capturing of water andnutrients, deep andintensiveroot architecture; abilitytointeract with beneficial soil microorganisms, like mycorrhizae and atmospheric N- fixing bacteria; efficient nutrient uptake and high nutrient use efficiency.Plant architecture for early soil cover and more light-competition. Good combining ability in variety and species mixtures.
Recovering ability	Good ability to recover from mechanical harrowing/cereals; high seedling vigour, fast and high germination percentage; deep and intensive root architecture.

Organic bakeries try to avoid synthetic additives in the bread-baking process, but then require varieties with high baking quality and still sufficient yield under lower input conditions. Another aspect of quality is taste. Sensory qualities, such as taste, are not only a result of environmental and management factors but also of genetic influences. Although sensory quality has been added to organic breeding programs of carrot, genetic variation for quality traits has not yet been fully explored.

Crop Ideotypes

Priorities for required traits differ for each crop and market sector. To improve the communication between farmers, traders and breeders, crop ideotypes should be elaborated and designed for each crop. In the short run such crop ideotypes can contribute to a more adequate selection of appropriate varieties among the existing assortment of conventional varieties. But they are also necessary to achieve selection criteria for new varieties from breeding programs specifically focused on low-input and Organic Farming. Organic crop ideotypes have many selection criteria in common with conventional required traits, such as disease resistance and high yielding ability. The difference is that varieties for organic farming have to be able to perform well under low-input conditions. Another difference is additional supporting characteristics for plant health and yields potential, such as long stems, the ears being high above the flag leaf, and the ears not too compact. But also, a characteristic such as a long stay green index for the top leaves in wheat is very important to enhance photosynthesis during the final phase of grain filling. It is important that the crop ideotypes have been designed with participation of farmers to include all practical experience as much as possible.

Variety Testing and Seed Production

In many countries, especially in Europe, trading of seeds of varieties of arable crops is regulated, which means that varieties should be registered on a national variety list, a European list, or a list of another country. To become listed, the variety should be tested using the so-called VCU protocol by a designated institution and shown to outperform existing varieties. Such a conventional protocol is designed for testing varieties under conventional farming conditions and could, therefore, impede the introduction of new varieties with specific low-input traits that are not included in such a protocol. Another critical aspect is that some traits required for organic farming systems are not expressed under conventional conditions, such as nutrient efficient use of organic fertilizers. Research organizations in some countries, for instance Austria, Germany, Switzerland and the Netherlands, have recently been successful in influencing the protocol for testing cereal varieties for the organic sector by including traits important for evaluation for organic farming systems and criteria to select research sites. Official authorities have adopted such a protocol, and they are engaged in a process to allow VCU testing for cereal varieties under organic conditions. As most of this VCU testing is being financed as a research project that aims to compare the results with those gained under conventional testing, the future of this testing is uncertain.

Seed Production

Organic seed production is still largely a missing link in the organic food production chain. Establishing organic seed production is also part of the process of developing new selection criteria for new varieties, from plant breeding, via maintenance to release and propagation of varieties. Although there have been many small-scale farmer initiatives to propagate varieties organically, large-scale use and production of organic seed production is still in its infancy. Some commercial companies have entered this market, but still not all existing or desired varieties are or can be propagated organically for technical or economic reasons. The two questions associated with organic seed production are:

1. Which varieties should be propagated; and
2. How should it be done?

The first question is largely answered in the former paragraph; the second one will be addressed by discussing diversity, technical aspects and seed quality standards.

Diversity of Varieties

In Europe and the USA, regulations no longer allow derogation for the use of conventionally propagated seeds of crops for which a sufficient assortment of organically propagated varieties and seed is already available. As this seed assortment usually consists of a minimum number of varieties, it has immediate consequences for the diversity of varieties that can be used in the organic sector and creates tension between short and long-term goals in the organic sector. There is the need not only to close the organic food supply chain as soon as possible to increase accountability for consumers, but also to have a diversity of varieties to meet the requirements of different market sectors and farming conditions. There are roughly two groups of farmers involved in this difficult area. The first is those small-scale farmers who focus on the local market with the use of local and old varieties, the so-called conservation or heritage varieties. Often, such farmers care for a community-based and participatory seed production and exchange system to conserve the local varieties or maintain old varieties that are suitable for low-input conditions. These farmers are threatened by restrictions from the new regulations because many of those conservation or local varieties are not officially registered and/or not propagated organically. When they want to save their own seeds from modern varieties, they have to deal with restrictive regulations on farmers rights for saving their own seed.

The second group is large-scale farmers who produce for supermarkets and have to meet specific quality and uniformity requirements. These farmers depend largely on modern, hybrid varieties of commercial seed companies and on the breeding policy of such companies. For economic reasons, not all companies are willing to enter the relatively small, organic market and produce all the desired varieties organically. Those who do take part in organic seed production produce only a limited assortment, which constrains the choice compared with the larger assortment of conventionally propagated varieties. During the first World Conference on Organic Seed, held in Rome 2004 (IFOAM 2004), it was recognized that there are large differences between countries with and without a large critical mass of organic products and organic seeds. It can, for instance, restrict seed export from developing countries, where the commercial viability of organic seed is lacking and seed certification is not existent, to countries in Europe with strict regulations on the use of organically propagated seeds. A careful transitional period for shifting from conventional to organic seed is necessary.

Technical Aspects

As organic farming systems refrain from the use of chemical protectants, it is even more important that seed and planting material is of high quality as this form the basis of crop production. The development of high-quality propagating material requires the development of specific expertise on aspects of seed production including technical knowledge, choice of location and varieties. The main problems in organic seed production are: nutrient management, disease and pest management, and weed control. Among the seed borne diseases in particular require special attention. To reduce the risk of disease infestation, optimal climatic conditions and thus the location for seed production can be important. In some cases, seed production should be located in warmer and drier climates, away from the areas of origin and destination. However, optimizing organic seed production not only requires adapting cultural practices as mentioned above, but also that specific attention needs to be paid for variety traits during seed production. This applies especially to biennial vegetable crops, which can build up disease pressure in the first year and then suffer from a continued increase of disease pressure during seed formation in the following year. Another aspect that influences the success of organic seed production is that some parental lines of hybrids have reduced growth vigour and, there- fore, are susceptible to biotic stresses including diseases. This implies that growth vigour as a variety characteristic is even more important in organic seed production than in conventional seed production.

Seed Quality Standards

Organic seed production without chemical inputs is a challenge for seed producers. The usual criteria for conventional seed quality also apply to organic seed: physical and genetic purity, absence of weed seeds, and a minimal requirement for germination. In some cases, it is not always possible to produce seeds without a certain degree of contamination with diseases. Several methods can be applied to improve the seed quality. One method is grading and separating infected seeds from healthy seeds by selecting based on seed weight or size, as known in the case of Fusarium on cereal seeds. For some seedborne diseases additional postharvest, non-chemical treatments are needed, such as hot water or hot air treatments. However, more research is needed to optimize such methods to reduce the risk of damaging the seed. Next to these physical treatments, disinfecting coatings with natural compounds are being developed, such as organic acids (mustard powder) or essential oils (thyme oil).

For some crops there is no problem in meeting the quality standards as required for conventional seeds, but in some cases the thresholds for seedborne diseases are adjusted. In some countries the recommended tolerances or thresholds for some diseases are lowered. In Austria, for instance, the threshold for Fusarium nivale has been adjusted from 20% in conventional agriculture to 10% for the organic sector. In the Netherlands, the level of permitted contamination in organic seed potato has been lowered from 25% (conventional) to 10%. A lower threshold is also set in other diseases, for example, in Austria, 10% is permitted for Septoria nodorum in cereals, compared with 20% previously.

Plant Breeding

Genetic Diversity

Modern varieties have been developed with the aim of combining high productivity and uniform product quality under high input conditions. The situation concerning the G.E.M interaction is different for organic farming systems and thus, the goal for breeding programs. First, under low-input conditions farmers have to cope with more biotic and abiotic stresses and larger environmental variation within and between farms, so the E is high. Second, an organic farmer has fewer means to overrule that variation, so the M is low. This has consequences for the genetic input (G) and the demands on varieties are therefore higher (see Yield stability and plant health). Organic farmers search for new varieties that possess a combination of required characteristics to increase yield stability under low-input, organic conditions. The question is how breeding can contribute to adapting crops and varieties to the organic, lower- input environment by improving the inherent buffering capacity of the farm system, and thereby, yield stability.

Modern varieties, such as those available for wheat, commonly lack the ability to adapt to different and changing environments, including drier regions with less fertile soils such as southern Australia. The tendency in conventional breeding is to select for single genotypes that have general adaptation to a large area where the variation in environments largely can be overruled by high inputs. Such varieties are generalists and are neither necessarily superior for nor adapted to a specific environment. As organic farming systems have fewer means to overrule agro diversity, the variation in environmental conditions is larger, and therefore the need for a larger variation of so-called specialist varieties exists. But developing specialist varieties is difficult to achieve economically. Second, organic varieties need a certain degree of buffering capacity, which means that the conventional pedigree line breeding may have its limits on the performance of crops, such as wheat, in Organic Farming. Next to that problem, the current new varieties are mostly derived from a limited number of parental lines and are thus genetically closely related to each other.

The genetic base for an organic breeding program should be broadened and newly established. This aspect of broadening the base is even more important when searching for adaptation to organic farming. A broader genetic base can be achieved by creating composite cross populations followed by selection under organic farming conditions. In composite crosses many selected varieties are intercrossed and all the hybrids are bulked together for propagation. Those selected crossing parents can consist of not only modern high-yielding varieties, but also varieties bred before 1960 (before the period of high inputs) that are capable of more efficient nutrient uptake under low-input conditions, and from old landraces as a source for adaptability. This approach can form the base for further selection of individual, superior genotypes and lines as potentially new and better adapted varieties. We can also expect to select individual lines with a good combining ability for variety mixtures. Variety mixtures formed by three or four existing varieties or extracted from composite cross populations are a simple and effective way of increasing the genetic diversity within a crop and improving yield stability. Mixtures can overcome several agronomic problems, through combining, for instance, high yielding types and types with a good baking quality and weed suppression in wheat. Research by conventional scientists indicates that there are areas of common interest to both organic and conventional plant breeders, for example, crop traits related to competitiveness against weeds and soil-borne disease prevention. However, any breeding strategy of restoring and enlarging functional genetic diversity requires reconsideration and adaptation of the legal and administrative frameworks. The regulations for registration are still based on the pure line practice. However, the market must be prepared to accept heterogeneous crops and products.

Conservation of Genetic Resources in situ and on Farms

In order to maintain greater genetic diversity in situ and a wide gene pool to improve genetic resources for the organic sector, seeds from the formal (institutionalized or commercial) and informal (farmer-based) seed systems should be accessible to local and national breeding programs. There is a pronounced need to identify appropriate genetic resources either for direct use or as potential parental lines in breeding programs. Evaluating and exploiting gene bank material can be of use because required characteristics might have disappeared by selection under modern, high input conditions, such as low-input tolerance and deep or intensive root architecture. This is a relatively low-cost method to develop valuable accessions.

Selection by Participation

As the formal seed sector cannot start breeding programs for limited acreage of Organic Farming easily, alternative options should be developed to enhance varietal diversity. Decentralized and farmers participatory approaches are known in developing countries and can offer an opportunity to combine knowledge and expertise of the organic farmers and that of formal breeders. Farmers can select in existing open-pollinating varieties or take part in the selection process of new crossings or developing base populations (see Genetic diversity).

There are different approaches to involving farmers in the breeding process. In the traditional farmers breeding model, farmers conduct the whole process of:

1. Selection of source germplasm;
2. Trait identification (prebreeding);
3. Cultivar development; and
4. Varietal evaluation.

In a complete participatory breeding model farmers are involved in all four activities in co- operation with breeders with a commercial or institutional background. On the other side there is the scientific breeding model completely conducted by the formal breeders. The efficient participatory breeding model, in which formal breeders involve farmers in the phase of selecting source germplasm and in the end phase of evaluating potential varieties; and the participatory varietal selection model, in which farmers deal only with varietal evaluation. In the Netherlands, there is a long tradition of cooperation between the formal potato breeding sector and farmer breeders. Farmer breeders with their daily practical experience and developed intuition (farmers eye) are involved in selecting the germplasm for new combinations together with the formal breeders. The formal breeder conducts the complex crossings between wild relatives and modern varieties, and offers the interested farmer breeders the seedlings for further selection during three years. The most promising phenotypes are delivered to the formal breeder or breeding company which will conduct further testing to see if it can end up in a variety. Also, organic farmers are involved in this participatory potato breeding system. Benefits derived from new varieties bred by farmers require a legal system of common ownership that allows equitable access and benefits sharing. In many countries there are examples of networks of organic farmers and breeders that provide such a system.

Genetic Engineering, in vitro Techniques, and Ethics

In ethical discussions about the application of modern biotechnology techniques, it is genetic engineering that attracts the most attention. A distinction is made between extrinsic arguments, dealing with the consequences of making genetically modified organisms (GMOs), and intrinsic arguments about the technology itself. There is in general much discussion about the meaning and validity of the intrinsic arguments, which are also dominant in the rejection of GMOs by the organic sector. In ethical committees and public debates, the emphasis is on the so-called extrinsic concerns: the risks for human health, for animals and for the environment. Most methods of risk analysis look only at the consequences and the effects of genetic engineering within the framework of a utilitarian ethics (weighing costs and benefits).

The intrinsic concerns deal with the technology itself rather than the consequences. Intrinsic objections are that genetic engineering is unnatural, or violating the dignity or integrity of plants and animals. These arguments are much more controversial, especially for those scientists and ethicists who consider extrinsic arguments to be more value neutral and objective. The intrinsic concerns, however, get more support from non-utilitarian ethicists, who refer to human virtues (virtue ethics). These virtues, such as humility and respect for (the intrinsic value of) nature or life, are important in people's basic attitudes towards nature. This ethics of integrity is related to a biocentrism ethical theory, in which all living organisms have a value (good) of their own, related to their species-specific or characteristic nature'. To say that a living organism has intrinsic value means that it has a value, which goes beyond its instrumental value for human beings. A living organism should never be reduced to a mere object for human use, as if it would be morally indifferent. See for an overview of these different theories. In the next section we will describe why biocentric ethical theory and intrinsic concerns are central elements in the rejection of genetic engineering by Organic Farming.

Why does organic farming oppose the use of GMOs?

The International Federation of Organic Farming Movements (IFOAM 2002) is opposed to genetic engineering in agriculture in view of the unprecedented danger it represents for the entire biosphere and the particular economic and environmental risks it poses for organic producers. The reasons mentioned by IFOAM can be clustered into three groups:

1. Risks for human health and the environment
   - negative and irreversible environmental impacts
   - release of organisms that have never before existed in nature and which cannot be recalled
   - pollution of off-farm organisms
   - unacceptable threats to human health.
2. Sociotechnical reasons
   - pollution of the gene pool of cultivated crops, microorganisms and animals

- denial of free choices, both for farmers and consumers
- violation of farmers fundamental property rights and endangerment of their economic independence.

3. Incompatibility with the principles of sustainable agriculture.

Human Health and Environmental Risks

The rejection of GMOs in Organic Farming is based more on a different perception of risks than upon the presence or absence of scientific proof that risks exist objectively. The perception of risk from an organic perspective is based on a holistic view of life. According to the IFOAM EU Group (2003), we also view this technology as inherently risky, because it is based on the reductionist scientific principles that have been shown to be flawed and are increasingly discredited. Some of the reasons why genetic engineering is considered to be a risky technology by Organic Farming proponents include:

- The gene constructs used are synthetic constructs. The introduction of pure DNA would lead to its rejection by the receiving organism. The production of synthetic constructs is needed to have any effect. Whether they work or not is tested in a trial-and-error process.
- The low efficiency (rate of success) of the technology.
- The introduction of foreign DNA leads to many unintended and unexpected effects.
- Genetic engineering (and the risk analysis, currently used by scientific committees) is influenced by the belief in genetic determinism, which is only part of the truth. Non- genetic, epigenetic influences during the development of the organism as a whole have shown to be as important as or even more important in some cases than the influence of DNA. The expression of DNA in the genome is much more dynamic than previously thought.
- Reductionist approaches to problems in agriculture are seen as symptomatic treatments. There is evidence of the resistance of insects to the pesticides used in combination with GMOs, or to Bacillus thuringiensis that is used in insect-resistant GMOs.

It is the dynamic complexity of the organism or ecosystem as a whole that leads to unpredictability and unintended effects. The stability of the gene constructs and the controllability of the technology cannot be guaranteed. From a holistic point of view, genetic engineering is based on a way of thinking that is characteristic of the physicochemical sciences. By formulating the question of risks in terms of holistic risk perception, the argument turns from an extrinsic one into an intrinsic one.

Incompatibility with Sustainable Agriculture Principles

This argument is problematic as long as these principles are not made explicit. Behind different meanings of sustainability are different views on the human relation to nature. The application of the industrial approach to living systems is felt to be unnatural in Organic Farming, not fitted to the nature of living systems. In the industrial approach, the tendency is towards full control over nature (anthropocentric attitude); nature in some sense is eliminated. Its opposite is pristine nature, that is, untouched by human beings. But it is impossible to speak about pristine nature in connection with agriculture, because every form of agriculture means interference in nature. One way to describe the position of Organic Farming is to say that nature and culture are seen as two poles of a polarity relation, and both poles have to be cared for. We could call this the integration of nature and culture or agriculture, but one that respects the independence, intrinsic value and autonomy of nature and of all living beings. This manifests itself in various ways, which lead to further arguments against genetic engineering:

- The use of natural, rather than synthetic substances. The gene constructs put into organisms to create GMOs are synthetic constructs, not natural substances. The most important difference with traditional breeding is the direct interference in the genome of plants and animals by forcefully introducing artificial and synthetic gene constructs (that can only be created in an artificial environment in vitro). These gene constructs are human inventions. And in general (non-evolutionary time scales), the gene constructs contain genes that would never be transferred by natural means (transgenesis).
- Stimulation of the self-regulation of organisms and the ecosystem (using natural processes). In Organic Farming genetic engineering is seen as a technology that forces the organisms to do what humans want, instead of eliciting a reaction in which the natural entity retains its relative independence as a partner. Illustrative is the way humans have dealt with reproduction in the process of domestication of cows (and other domestic animals). Step by step (artificial selection – artificial insemination – embryo transplantation– genetic modification and cloning) the animal's own role in reproduction, its independence, is taken away from it and brought under human control.
- Respect for the specific characteristics (nature) or intrinsic value of plant and animal species, (agro-) ecosystems and landscapes. This is another reference to the acknowledgement of the independence of nature, but now at a moral level. In this moral context also the word integrity is used. Explained that the integrity of plants refers to their nature or way of being, their wholeness, completeness, their species-specific characteristics and their being in balance with the species-

specific environment. When integrity refers to the specific nature of plants (or animals), different levels of integrity can be distinguished: integrity of life, plant-specific integrity, genotypic integrity and phenotypic integrity. With this moral instrument the authors have assessed different plant breeding and propagation techniques, to find out whether they respect the integrity of plants. The outcome is that techniques at DNA level (e.g., protoplast fusion, genetic modification) violate all levels of the nature (integrity) of plants. The conclusion is that genetic engineering is not compatible with the intrinsic value of plants.

Our conclusion is that intrinsic arguments against genetic engineering are widely used in the field of Organic Farming. Such arguments usually rest upon a specific view of the human–nature relationship, which includes cognitive, emotive and volitional elements. Cognitive elements refer to a holistic (non-reductionist) view of living organisms. Emotive elements refer to a biocentric attitude towards life in which living organisms are seen as partners that should be respected, with an intrinsic value. The volitional elements refer to ethical statements as to what should or should not be done in Organic Farming, taking other elements into account.

Respecting the integrity of plants can also have consequences for the application of other modern breeding and propagation techniques for organic breeding and propagation programs. It may imply that:

- reproductive barriers between species will be respected and not violated;
- in vitro techniques are not compatible with organic principles;
- sterility, such as cytoplasmic male sterility, will not be accepted in the end product (variety) without including restorer genes; and
- patents on life are not accepted.

These considerations are not yet implemented in the Basic Standards for Organic Production and Processing of the International Federation of Organic Farming Movements and have only appeared as draft standards for 2002-2005.

Molecular markers in use in Organic Breeding Initiatives

DNA diagnostic techniques, which enable selection at DNA level, do not involve genetic modification. The techniques, which are usually based on biochemical and molecular markers, could therefore be used in organic breeding programs to supplement trait selection methods in the field, but their potential for Organic Farming has yet to be proven. Genotype–environment interaction is of primary importance in organic plant breeding, and markers can, in addition to field selection, contribute to assessing environment-specific performance of genotype traits. However, a point of attention is the techniques used with DNA diagnostics, as some of them include the use of radioactive isotopes and (cancer-inducing) chemicals, which are not used or permitted in Organic Farming.

Obstacles to Organic Farming

While Organic Farming aims to be environmentally sustainable, it has not yet reached its goals and there are issues that still need to be addressed. Many of these issues are reviewed in detail in other chapters of this book, including several key topics that were selected for particular analysis, such as the impact of tillage in Organic Farming and the industrialization of organic production systems. A common question asked of the organic movement relates to its yields:

- Can Organic Farming feed the world? Like questions about sustainability, productivity also depends on many factors including the farmers background, the farms resourcefulness and local and national support mechanisms.
- Conventional agriculture successfully feed the world now? High input-high yielding systems are currently failing to feed the world, not because of problems with productivity, but because of problems with food distribution and social organization, and serious concerns such poverty, racism and gender imbalance.
- Comparisons of organic and conventional farming have been a common feature of the organic literature since the 1980s.
- The researchers have looked at a wide range of measures including yield, economics, resource use efficiency, environmental impacts and social factors on a diverse range of farm types such as dairies, orchards and mixed cropping farms.
- Some important examples of comparative research have been published in prestigious journals, providing valuable credibility for claims that Organic Farming is productive and sustainable.
- Additionally, numerous other studies have been published in academic journals of various disciplines. Some key findings from research that has examined yields suggest trends including.
- Challenges for Organic Farming while Organic Farming aims to be environmentally sustainable, it has not yet reached its goals and there are issues that still need to be addressed.
- Many of these issues are reviewed in detail in other chapters of this book, including several key topics that were selected for particular analysis, such as the impact of tillage in Organic Farming and the industrialization of organic production systems.

- A common question asked of the organic movement relates to its yields: can Organic Farming feed the world? Like questions about sustainability, productivity also depends on many factors including the farmers background, the farm's resourcefulness and local and national support mechanisms.
- The appropriate answer may be: does conventional agriculture successfully feed the world now? High input-high yielding systems are currently failing to feed the world, not because of problems with productivity, but because of problems with food distribution and social organization, and serious concerns such poverty, racism and gender imbalance.
- Comparisons of organic and conventional farming have been a common feature of the organic literature since the 1980s.
- The researchers have looked at a wide range of measures including yield, economics, resource use efficiency, environmental impacts and social factors on a diverse range of farm types such as dairies, orchards and mixed cropping farms.
- Some important examples of comparative research have been published in prestigious journals, providing valuable credibility for claims that Organic Farming is productive and sustainable.
- Additionally, numerous other studies have been published in academic journals of various disciplines. Some key findings from research that has examined yields suggest trends including.

Farming systems comparisons, preferably conducted over several years, supply valuable information about agricultural productivity and performance. However, they are subject to important limitations including management×site×variety interactions and externalities (e.g., energy, pollution and health) that may not be taken into account. High levels of government and commercial support have been invested over many decades in optimizing plant and animal germplasm, soil fertility and pest management systems, and human capacity for conventional farming systems. This support would be expected to create substantial advantages for conventional producers. Research methods for comparative systems trials are continually being refined, not only regarding agricultural and ecological considerations, but also social and statistical issues. In addition to productivity, the importance of other farming systems attributes such as resilience and stability have also been highlighted. Organic maize out yielded conventional maize by significant margins in 4 out of 5 drought-affected years. A range of new frameworks are being developed for addressing externalities, environmental impacts, labour relations and so on. These frameworks include EMS, input-output analysis and life cycle analysis.

Protection of Crops in Organic Agriculture

Organic farming systems are challenged by many of the same crop protection issues as conventional farming systems. Approaches to crop protection in Organic Farming differ widely among growers globally and regionally. At one end of the spectrum, organic growers use substitution-based approaches in large-scale operations to capture premium prices in a niche market. At the other end, resource-poor farmers producing subsistence crops use, by default, pest regulation tactics based on traditional knowledge. Organic growers at both ends of the spectrum are less motivated by environmental and public health considerations than are those growers that have formed the philosophical center of Organic Farming movements in various parts of the world. For these growers, Organic Farming differs fundamentally from conventional agriculture, not in terms of the pest and disease challenges that face crop production or solely in the range of tactics used by growers, but in the conceptual approaches that frame crop management strategies.

Too often, descriptions of the conceptual approaches in conventional and Organic Farming are overly simplified. Conventional pest control can no longer be characterized as the reliance on scheduled applications of broad-spectrum pesticides (biocides, insecticides, fungicides, herbicides). Best practices in conventional agriculture incorporate a wide range of tactics, including pest monitoring and judicious use and timing of selective pesticides, selection of insect and disease-resistant cultivars, and cultural controls such as crop rotation and crop residue destruction. By the same token, Organic Farming is more than conventional agriculture minus synthetic fertilizer and pesticide inputs. While some organic growers do simply substitute manure for fertilizer and botanically derived pesticides for synthetic pesticides-more often, organic practices involve a wide range of soil management and cropping practices that maintain ecosystem health and foster ecosystem services.

For the purpose of this chapter on pest and disease management, Organic Farming is defined as plant and animal production systems managed with an emphasis on sustainable and renewable biological processes: nutrients supplied at rates needed to maintain nutrient balances through decomposition of nitrogen (N)-fixing green manures and plant or animal-based soil amendments, and pest management relying heavily on promoting plant health, vegetation management and biological control. Curative pest treatments include application of microbial, botanicals, soaps, oils and minerals and augmentative releases of predators; synthetic fertilizers or pesticides are generally not applied, unless exemptions are granted.

We will use a theoretical approach to the characterization of pests and diseases in agricultural systems, based on some of the ideas from invasion ecology. Invasion ecology is a complex and dynamic intellectual conversation often focused on exotic plants, mammals and birds. Here, we don't consider only exotic species as invaders but also any pathogen or pest species not yet present in a crop in a particular growing season. To apply some basic concepts of invasion ecology to crop protection, we consider three phases of invasion:

1. Colonization;
2. Establishment; and
3. Population outbreak.

We explore whether certain invasion trends may be distinguished for herbivorous arthropods, nematodes, fungi, and bacteria in organic versus conventional farming systems. By borrowing the term invisibility, we can examine how organically managed crops may present barriers against the invasion by pests and pathogens, and compare them, when possible, with conventionally managed crops. In natural ecosystems, low invisibility has often been related to high biodiversity depending on the scale of observation, and this relationship may also hold for managed agro ecosystems. Organically managed agro ecosystems are generally more diverse than their conventional counterparts. This has been shown for above-ground natural plant and crop species, insects and birds, as well as for below-ground arthropods, nematodes, fungi and bacteria. The reasons for this difference in biodiversity are manifold, but in particular, the:

- Absence of herbicides reduces detrimental effects on various microbial species;
- Absence of synthetic nematicides and insecticides reduces broad-spectrum effects on beneficial fauna;
- Absence of general fumigants reduces broad-spectrum activity on all soil life;
- Absence of easily available plant nutrients reduces the selective enhancement of fast-growing microorganisms; and
- Addition of various plant and animal-derived organic materials enhances the soil food web and, indirectly, the above-ground food web.

In addition, organic farmers frequently purposefully plant strips of controlled natural vegetation, which affects not only above-ground biodiversity, but also soil biodiversity. If higher biodiversity in agro ecosystems reduces invisibility, then we can expect a reduced spread of pests and diseases in organic compared to conventional farms. We suggest that this expectation is largely met, but that there are exceptions. We compare the range of pests and diseases that challenge crop productivity in organic and conventional farming systems around the world. Our emphasis here is on arthropod pests and diseases, but we include some observations on vertebrate and other invertebrate pests. We then describe organic and conventional strategies for pest and disease management, including the three elements:

1. Prevention of colonization or establishment;
2. Population regulation through biological processes; and
3. Curative interventions.

To illustrate these elements of pest and disease management in practice, we critically review comparative research programs on conventional and organic pest control in many parts of the world, and illustrate the constraints and opportunities of organic crop protection under different farming conditions. We conclude by providing suggestions on future research directions that will advance our knowledge and capacity for effective crop protection in Organic Farming.

8

A Human's part in Organic Animal Herds *Health Promotion*

In regard to health and wellbeing, knowledge of the behavioral patterns of the animals is important, as well as the intense surveillance of the animals and the herd combined with immediate intervention when necessary. In some systems with large flocks of animals, for example in Australian beef herds, the individual animal may be living a highly natural life, but it is not as closely looked after as an individual animal in more intensive production systems. In the latter case, closer proximity means the farm operator is more likely to observe problems and intervene early, reducing possible prolonged suffering.

Treatment of Animal Diseases in Organic Animal Husbandry

Intervention in any case of potential crisis: critically important, one aim of organic farming is to reduce the use of synthetic chemical substances. However, disease must be treated for animal welfare reasons, and the most effective treatment will often be considered to be biomedical veterinary medicine of synthetic chemical origin. The compromise in European countries regarding the use of synthetic chemical veterinary medicine, is to encourage health promotion and non-medical disease prevention by demanding a two to three times prolonged withdrawal time after treatment, and a limited number of treatments per animal per year or lactation. The farmer has a responsibility to act and intervene appropriately in relation to each individual animal in the case of disease. It is not a question of whether to intervene, but about which form of treatment, if any, will best prevent suffering and aid recovery. Instead of thinking only of disease treatment, options for supportive treatments should be considered. These comprise ways of supporting the animal in an unbalanced health state to recover. Taking dairy production as an example, a solution could be milking by hand between machine milking very early in a mastitis case; providing extra bedding in critical situations (e.g., after calving), using soap and water to clean infected wounds, traumas or some claw disorders (e.g. abscesses). Longer term strategies include culling, drying off single udder glands, and letting cows with high somatic cell counts stay together with suckling calves for a short or long period. Careful inspection of animals and their products (e.g., milk) is of crucial importance, and immediate reaction to any critical condition is the most important factor. Many of these interventions have relatively high labor demands, such as udder massage and milking by hand, which may limit their relevance in, for example, intensive northwest European farming systems. Previous and ongoing Danish studies show that the basic ideas of organic farming can be seen as stimulating a great effort to improve the health situation in herds, including disease management by farmers. In interviews, many organic farmers have described a 'conversion in the herd and heart', which made them take responsibility for their own animals in a new way instead of just calling the veterinarian.

Alternatives to the Standard Techniques of Treating Disease

In the US organic livestock standards, there is a complete prohibition of biomedical product use, which emphasizes the need for the development of sustainable treatment methods for organic animals. Phytotherapy represents a wide range of different methods and substances primarily based on plant products, and will in many cases be defined within the framework of folk medicine. Homoeopathy is a medical school with its own theory and practice, and presents an approach to health and disease very different from that of conventional biomedicine. In several European countries, there has been a discussion of the legal status of phytotherapeutic as well as homoeopathic products, as they are not registered for use in production animals, which forms a bureaucratic restriction on the development of these methods. Development of various research frameworks and study designs have been discussed, to create a better understanding of the mechanisms of this treatment method. Acupuncture and chiropractic are other options for the treatment of production animals. As with any treatment method, they will demand a relatively high level of specific education. Acupuncture treatment will often require a veterinarian, and sometimes a treatment will take 20-60 minutes, which limits its use in many situations. In production animals, acupuncture has been shown to be particularly useful in reproduction disorders of all kinds including birth situations.

The Participation of the Veterinarians

In relation to health promotion and disease prevention as well as disease management, the role of the veterinarian and other animal health professional advisers should be to support the farmers in fulfilling their goals for the herd as well as organic goals. In the UK, health planning is a part of conversion to organic farming and in Switzerland, a research and development project involving health planning in several organic dairy herds has pointed to the importance of the dialogue between farmers and advisers to develop organic strategies for better health and less disease. In Denmark, more studies have shown that farmers, to a small extent only, involve their veterinarian in the development of the organic herd, and that one major reason is a lack of understanding and support from the veterinarians, who do not consider 'organic farming' to be something special that they should relate to. In a working group report at a European workshop, those veterinarians seemed to have a general lack of knowledge and to be sceptical of organic farming. Consequently, a great potential seems to exist in terms of development of collaboration on strategies for health promotion, methods of alternative disease management, as well as ways of interacting for farmers and animal health professionals. One recent and apparently very fruitful development in the organic environment seems to be the adaptation of the so-called Farmer Field School approach from African and Asian countries to Danish organic dairy herds. The formation of farmer groups, where farmers advise fellow farmers, seems to support the search for innovative solutions to health and disease problems, and thereby improves the health situation on the farm.

From organic livestock production to organic animal food production: the whole food chain Organic food: process quality and product quality

Two quality criteria can be used in organic livestock production:

1. Process quality, a description of the way the food has been produced (e.g., animal welfare, environmental care); the things consumers 'know' about the product and which add a certain quality to it, influencing their choice of this product; and
2. Product quality, which covers nutritional and sensory qualities, food safety qualities and possible health benefits.

Both aspects are important for organic food products, and both can be improved at the farm level through feeding and animal welfare related routines. Concluded that there are no consistent differences in nutritional qualities between organic and conventional animal products, and no clear evidence that organic food contains more mycotoxins or is more or less microbiologically safe than conventional food. Consumers are also motivated to buy organic food to avoid genetically modified organisms and food irradiation.

Aspects of Food Safety and Zoonotic Diseases

Several infectious agents can be transmitted from farm animals to humans through the food chain and cause disease (zoonoses). These diseases are often associated with highly industrialized livestock production, but they also can occur in organic production. In particular, the risk for zoonotic infections with outdoor reservoirs, such as Campylobacter (in poultry), is increased. Another important zoonotic infection is Salmonella, although little difference in incidence of this infection is reported between organic and conventional production, in both poultry and pigs. However, several factors may reduce the risk: different feeding regime, lower stocking rate, grazing, feeding roughage (e.g., silage with a low pH). Salmonella bacteria may survive outdoors for up to one year, and therefore the resting period of pastures may be important. For example, found 35% of organic poultry sampled positive in the Netherlands for Campylobacter and 13% for Salmonella. The measures for prevention of certain diseases and the risk factors for food contamination like Salmonella can be considered as premature. Trichinosis is an example of a zoonotic helminth infection that may have a reservoir in wild pigs and foxes. The risk that such an infection is introduced into outdoor organic pig herds is substantially higher than in indoor production, likewise toxoplasmosis Bovine spongiform encephalopathy (BSE) is now regarded as a probable zoonosis. The disease was spread by recycling meat and bone meal made from infected sheep or cattle as feed for livestock, and had originated in the early 1970s from a novel source, possibly a cow or other animal that had developed the disease as a result of a genetic mutation. Since BSE has been related to the consumption of bone meal, and this has not always been prohibited in non-organic herds, there might be a risk of BSE in animals that have been imported from conventional herds.

Biosecurity and Animal Management-related Risks

Certain production conditions present particular biosecurity risks. For example, where disease incubation or weed dispersal are difficult to control such as in intensive dairy systems or hill farming in the UK, there will be a greater likelihood of harmful organisms proliferating. The movement of animals between farms, often over long distances, adds additional biosecurity and animal welfare concerns. In many European countries animals for slaughter are transported from farms in remote areas to abattoirs, some of which are organically certified, or they are transported between farms and grazing areas. When discussing the potential for development of organic production in areas where this is not yet established, the export and import of animals, the transhumance and the use of communal grazing areas should be profoundly analyzed and managed in the context of each specific region.

Perspectives for the Future

Several challenges exist in animal health for organic livestock production, both in established organic farms and in areas where organic farming is developing. The prevalence of disease is still critical in many organic farming systems, and sustainable ways of managing organic herds still need further development. One major challenge for a more global development of organic farming is to further infiltrate the concept of 'alternative systems' into the many diverse existing farming systems. For example, communal grazing is a way of livestock production commonly practiced in large areas of the world, and this needs to be considered when converting to organic farming. It may involve conversion on the 'communal level' rather than the farm level in some cases. Many farming communities are extremely resource poor. Even though there might not be a particular market for 'organic products' from or in these communities, organic practices still may be relevant to support both environmental and economic sustainability of the area; this may include using local resources and not relying on import of feed, drugs and animal breeds. They may serve as local beacons for a sustainable development.

Disease management should be based on health promotion efforts and disease prevention strategies. When disease occurs, rapid intervention using alternative strategies and disease treatment methods if possible is critical. Veterinary services and extension need to be 'converted' to a much greater extent to support the organic approach to animal health and welfare. The role of humans in the organic herds is to both support the possibilities for the animals to perform their species-specific natural behavior, giving them as many opportunities as possible to live a 'natural life' (e.g., flock life, outdoor life, mother–offspring relations), and to be care-takers who support animal wellbeing by fulfilling their nutritional requirements and who intervene whenever necessary to avoid a potential crisis.

With regard to organic feeding, many of the challenges regarding home-grown or locally grown, organic feed can be met by optimal crop rotation systems, high quality feed production in terms of pasture management and harvest at the right time, choice of the best crops, conservation and storage under optimal conditions. However, some import into the farm seems necessary, such as trace elements and minerals. The reliance by monogastric animals on synthetic amino acids in conventional farming systems needs to be met in organic systems using more sustainable, long-term strategies, including breeding and production goals. Differences between countries and regions with regard to farming systems seem bigger than the general difference between conventional and organic livestock systems.

9

Organic Foods Market

By 2002, the international market in organic foods and beverages was estimated to be worth US$23 billion. This brought the market share of certified organic food in the developed countries where it is predominantly consumed to somewhere between 1% and 2% of total food sales. According to critics, this small percentage proves organics to be little more than an overhyped Western food fad. Yet with ongoing sales growth in developed countries of between 8% and 20%, organic foods have attracted the interest of a growing number of farmers, food processing firms, retailers and governments. Along with fair trade goods, organics has become one of the fastest growing sectors of a global food market characterized more generally by oversupply and falling terms of trade. Indeed, it has been estimated that the international market in organic foods could reach US$100 billion as early as 2006 with the organic sector in the United States of America (USA) alone worth over US$30 billion by 2007. From an insignificant niche market as recently as the mid-1990s, organics has leapt into the mainstream.

This is not to say that this growth will continue indefinitely. The expansion of larger European markets appears to be slowing as they approach what some commentators refer to as maturity. And as expansion slows, price premiums come under pressure, as reflected in a dramatic fall in the farm gate prices paid to organic dairy farmers in the United Kingdom (UK) in 2001 as a result of apparent oversupply. In other markets, the issue appears to be less one of demand slowing than of producers failing to supply produce of sufficient consistency in quantity and quality to secure a sound distribution and retail base. Despite conditions of apparent under-supply, around 35% of certified organic produce grown in Australia is sold as conventional while 40% of organic beef and 25% of organic milk in

Belgium is sold as conventional. Compounding these demand and supply-side constraints on the continued expansion of the organic market is the simple issue that the organic industry knows little about who purchases its products or why they purchase them. Common lore tells us that organic market growth has been rooted in an unlikely mix of fear and fashion. Many consumers, it is believed, have been attracted to organics by their dread of the imperceptible threats of industrial food production. The mad cow crisis and other food scares have so tested consumer confidence in conventional food systems, and the institutions that regulate them, that assurances regarding the safety of genetically modified organisms, food irradiation and agricultural chemicals have fallen on deaf ears. Others, purportedly, have been attracted by the positioning of organics as a premium quality niche product. Substantially higher retail prices for organic foods have promoted a perception that they are consumed primarily by those wealthy enough to pay virtually any price to purchase what they believe to be tasty, safe or nutritious foods. There is certainly a grain of truth in these beliefs. However, research does show organic consumption to be somewhat more complicated.

In this chapter we examine the organic marketplace in terms of the changing dynamics of supply and demand; what is known about those who consume organic foods; what strategies have been set in place to recruit more people as organic food consumers; and constraints and opportunities for further expansion of the market for organic foods. In doing so, we will address, at least in part, one of the principal concerns that has arisen within the organic sector as a consequence of its growth; that is, the concern that expansion has come at the cost of the core values and practices on which Organic Farming was founded.

The Demand for Organic Foods

There is a widespread perception that rapid growth in the organic market since the mid-1990s has been based more on an escalation of consumer demand than on an expansion of supply. This demand–pull perspective on market growth is lent credence by the claims of several distributors, processors and retailers that they have experienced considerable difficulty sourcing enough certified produce to satisfy buyers, leading some to sponsor the conversion of conventional farmers to organic production. Yet, as we have seen, there already are signs that recent rates of market expansion are slowing. Demand, however, is a difficult concept to quantify. There are few people who, when asked, will state categorically that they would not purchase organic foods under any circumstances. Surveys suggest that were organic foods to be made as widely available as conventional foods; almost all Western consumers would buy them at least some of the time. And the lower any retail price premium is over outwardly similar conventional foods, not surprisingly, the more people express an interest in purchasing organic produce. The level of demand, therefore, for organic foods derives not only from the values, beliefs and socioeconomic status of food consumers, but on the relationship between these attributes and supply side issues such as quality, distribution and retail pricing. Levels of demand, even within markets that appeared to have reached maturity, may shift very rapidly in response to major changes in supply. Given these factors, demand is examined in this section, from the perspective of existing supply side dynamics, retail price premiums and consumers stated willingness to pay for certified organic foods.

Generating an accurate picture of the international organic sector accurately is hampered by limited data availability and uniformity. Nevertheless, the most recent International Federation of Organic Farming Movements report suggests that of the US$23 billion global retail market for organic food and drink in 2002, North America accounted for US$11.75 billion, Europe US$10.5 billion, Japan US$350 million, Oceania US$200 million, Latin America US$100 million and the rest of Asia and the whole of Africa less than US$200 million. These figures reflected growth, in 2002, of about 12% in North America, 8% in Europe, and 15% to 20% in Australia. Following several years of growth of between 20% and 40%, the UK most exemplified the slowing of market expansion in those European countries with comparatively large organic markets. Importantly, if this thumbnail sketch of the international organic marketplace is compared with broad international patterns of organic production, there are some notable points of divergence. Of the roughly 24 million hectares managed worldwide for certified organic production in 2002, over 10 million were located in Australia, 5.8 million in Latin America, 5.5 million in Europe, 1.5 million in North America, 880,000 in Asia, and 320,000 in Africa. Despite the spatial dominance in Australia and Latin America of semi-arid rangelands used for extensive cattle grazing, it is strikingly apparent that the poor countries of Africa, Asia and South America are producing significant quantities of organic foods almost exclusively for sale in Europe and North America. The same is true within Europe, where Italy and Spain act mostly as exporters to Germany and the UK, which imports around 50% and 65%, respectively, of their organic food needs.

Average national retail price premiums for certified organic food have been reported at 20% to 30% in Austria, 10% to 15% in Germany, 10% to 100% in both the USA and UK, and 80% in Australia. Considerable variability also may be found both between commodities and within countries, with those products that are supplied in bulk, and widely distributed to consumers, tending to receive the smallest price premiums.

This is the case for organic milk in both the EU, where retail premiums on milk are small to non-existent, and Australia, where they are about 35%. In Denmark, where consumers display the second highest rate of expenditure on organic foods in Europe after Switzerland, retail price premiums are reported to be negligible. The general pattern, however, is of retail premiums remaining considerably higher than the 15% to 20% that most surveys suggest is acceptable to most Western consumers. That a retail premium of 25% to 30% is acceptable to enough European consumers that retailers may realistically charge this much while still broadening their organic customer base, existing retail premiums would seem a significant limitation on the ultimate size of the organic market. Before concluding, however, that retail price premiums alone will ensure that organics never grows beyond its status as a lucrative niche market; additional points need to be considered. First, it is impossible to predict the impact that future events and technological developments will have on consumer trust, or distrust, in conventional food production and regulatory systems. Another mad cow crisis may well see perceptions regarding the acceptability of different price premiums revised upwards. Conversely, serious organic food scares may challenge perceptions that organic foods offer a safe alternative. The percentage increase in the price of individual organic foods that is important to most consumers but the absolute price increase. In other words, most consumers may be willing to pay a higher premium, in percentage terms, for less expensive goods on the assumption that this will make little difference to their overall grocery bill.

Why individuals like to eat organic foods or not

International research suggests that those attributes of organic products most likely to influence consumers are, from most to least important:

1. Health (i.e., minimal artificial chemical residues in the product and high nutritional value);
2. Environment (i.e., environmentally friendly production and processing);
3. Taste;
4. Animal welfare;
5. Minimal processing;
6. Novelty; and
7. Fashion.

Conversely, those attributes most likely to limit consumption of organics are also identified from most to least important:

1. High price;
2. Limited availability;

3. Scepticism about the credibility of product claims;
4. Poor appearance;
5. Non-awareness of organic; and
6. Contentment with existing products.

According to, the typical organic consumer is urban, well-educated, from middle to high income household, and discerning in their food choices. Market researchers in the USA posit a more differentiated mix of organic consumers that includes: a small group of passionate environmentalists; an older, wealthier group concerned mostly about their own health; a young group who profess environmental concern but who tend to act on that concern only when convenient; and a growing mainstream who care about the environment and are willing to engage in green consumption as products become more accessible. Italian researchers, however, differentiate organic consumers into the health conscious (the largest group), the curious (and open to more sales), the pragmatic (who are very concerned about higher prices) and the nostalgic (who associate organics with the tastes and authenticity of the past). Each of these claims seems to make sense in light of the high retail premiums paid for organic foods. But it is important not to dismiss organic consumption as the domain only of the rich and passionate. Most people who have no immediate interest in green consumption are those who struggle for economic survival, an underclass of the unemployed, underemployed and underpaid. If this group of very poor are taken out of consideration for a moment, some 75% of consumers in the USA emerge as genuinely interested in purchasing organic foods.

The belief that food should be safe, nutritious, tasty and environmentally responsible is not radical or marginal, so it should come as no surprise that many food consumers in Western countries profess to consume at least some organic food, 40% in Australia. Consequently, several studies from around the world have found that beyond the exclusion of the very poor there are few meaningful demographic differences in Western countries between those people who consume organic foods and those who do not. The most important dimension of demographic difference is not education or income, but gender, with significantly more women than men claiming to purchase organic foods. This may stem from the higher levels of responsibility taken by women for feeding children and other family members although; such responsibility can also place pressure on women to priorities convenience and price. Although many Western consumers claim to consume organic foods, and many more profess the values that underpin Organic Farming, overall levels of organic food consumption remain only 1% to 2% of total food sales. Estimated that committed organic consumers, those who claimed to consume half or more of their diet as organic, accounted for about half of all certified organic food sales. This group were more motivated than were occasional organic consumers to source foods they believed were natural (i.e., free from artificial additives, unnecessary processing, genetic engineering, irradiation, pesticides, preservatives, animal growth hormones, antibiotics), environmentally friendly, supportive of animal welfare, and likely to make them feel good emotionally. These groups did not express fundamentally different food values. Even those who did not consume any organic foods claimed to be motivated in their food choices by health, environmental and animal welfare concerns. The difference among the groups was how strongly these motivations were expressed relative to price and convenience. Examined how the motivations behind food choice interacted to influence increasing levels of organic food consumption. They found the major direct influence on increasing rates of organic food consumption to be consumers commitment to the consumption of foods they perceived to be natural. Women, and those responsible for household food provisioning, were far more likely than others to be motivated by this concern. This was followed by willingness to pay a premium for environmental values. The next most important factor directly influencing rates of organic consumption was the level of motivation towards the consumption of food that made the respondent feel good, physically and emotionally. Organic food was not only believed to taste and smell better, but also to evoke feelings of safety and tradition. Again, responsibility for household food provisioning and gender were the major determinants of consumers level of motivation towards sensory and emotional appeal.

Advertising of Organic Food

One of the things that are truly remarkable about organic foods, given their status as one of the fastest growing sectors of the food industry, is how little their marketing depends either on consumer research or on aggressive advertising and discounting. At face value, the demand-pull perspective on organic sector growth would suggest that organic growers have been in the enviable position of being able to concentrate their energies on expanding their farming operations while receiving premium prices for their produce. Despite a paucity of organic food advertising in the mainstream media, that media portrays organics as the almost sole alternative to environmental and food safety risks associated with industrial agriculture. Analysis of newspaper references to organic food and agriculture in the UK, USA and Australia suggests that the term organic has come to signify a loosely defined bundle of desirable attributes related to quality, safety, ecology, tradition and provenance. In a world where regulatory agencies, and the complicated systems of quality assurance they administer, seem incapable of guaranteeing the safety of conventional foods, the organic label offers a simple, recognizable and, for some, comforting alternative. Unfortunately, for organic growers, organic produce does not necessarily walk off supermarket shelves. That a significant proportion of certified organic produce is still sold on conventional markets demonstrates there is no guarantee that the use of organic methods will result in the sale of a clearly differentiated organic product. As a consequence, marketing efforts within the organic production sector has concentrated on the development of supply chains (i.e. on the development of distribution and retailing arrangements) and on the development of an appropriate regulatory regime to police usage of the term organic.

The single most important strategy in the marketing of organic foods has been the establishment of regulatory systems to oversee the development of standards and inspection systems for organic production, processing and labeling. In terms of market expansion, independent third-party certification of compliance with organic process standards achieves several things. Most obviously, certification provides some measure of guarantee for the buyer that they are getting what they pay for. As progress continues towards the harmonization of standards on both national and international levels, buyers may extend their confidence to produce sourced from almost anywhere in the world. Further, even though no set of standards can be expected to codify adequately all the principles of Organic Farming, or all the conditions under which it might be practiced, compliance with these standards implies, nevertheless, an integrity that encourages consumers to associate several desirable attributes with certified organic foods even though, strictly speaking, those attributes fall outside the scope of what is guaranteed. Such attributes include taste, healthfulness, and so on. From the grower's perspective, independent certification enables those who have undertaken it to make claims that, in many circumstances, have clear value in the marketplace. The flipside to this is that the regulatory regimes underpinning certification preclude those who have not been certified from labeling their produce as organic.

Argues that organic certification creates conditions of scarcity, by limiting the amount of food labeled as organic in the marketplace, and thus provides the basis for farm gate and retail price premiums. While notions of scarcity and price premiums intuitively would suggest that certification may limit market expansion, the effects may be both positive and negative. The cost and complexity of certification certainly does discriminate against smaller growers, particularly those in the developing world. However, scarcity and price premiums have acted to encourage those farmers and others who could afford the cost of certification into the organic sector and have thus helped to boost organic food supply and availability. This has been clearly demonstrated in the USA, where the implementation of a uniform national standard in 2002 is credited with boosting consumer awareness and confidence in organic produce and with clearing the way for its entry into mainstream supermarkets.

There are now few developed countries where mainstream supermarkets and other large retail chains do not appear set to dominate sales of organic foods. Among the world's three largest national organic markets, for example, supermarkets claimed 49% of organic retail sales in 2001–02 in the USA, 40% in Germany and 80% in the UK. Four years previously, the supermarket share of organic retail sales in Germany had been a mere 26%. In the USA, 31% of the organic retail market in 2001-02 was held by just three natural food retail chains (i.e. Whole Foods Market, Trader Joe's and Wild Oats Markets) that operated across multiple locations using store layouts and scales similar to those of conventional supermarkets. Reflecting the assumption, whether accurate or not, that organic consumers are urban, educated and comparatively wealthy, European supermarket chains have been shown to concentrate the availability of organic foods in larger, higher quality urban stores.

The increasing involvement of very large businesses, such as supermarket chains, in the organic sector has attracted concern and criticism. The argument is that in relatively small scale and localized food networks, enough opportunities are afforded to consumers for direct interaction with growers to ensure their confidence in the integrity of the foods they purchase. By replacing direct interaction, and the trust it encourages, with codified sets of standards, the way is cleared for organic foods to be shipped around the world at enormous cost in fossil fuels and at the expense of values that are not easily codified such as agricultural biodiversity and community building. The conventionalization thesis, as this has become known, assumes that the entry of larger businesses into the organic sector first displaces the smaller businesses that pioneered the sector and second, that these businesses do not share the values, nor implement organic standards with same authenticity, as organic pioneers.

There can be little doubt, however, that the increased visibility and availability of organic foods facilitated by mainstream supermarkets has been a major factor in expansion of the organic market. The reduced distribution costs and economies of scale enabled by the involvement of larger retailers have contributed to lower retail price premiums, and hence increased consumer demand, in those European countries where supermarkets dominate organic sales. Further, because the total volume of organic food sales has increased alongside mainstream supermarket involvement in the industry, the increased market share captured by supermarkets has not come directly at the cost of absolute sales among their competitors. Indeed, where small retailers, such as health food stores, have experienced declining organic food sales they have done so less because of mainstream supermarket competition than because of competition from specialist organic and natural food chains.

The other area of growth in organic food retailing (in absolute if not relative terms) has been direct farmer to consumer sales. Farmers markets, box schemes (i.e., local, not-for-profit groups buying and distributing organic food cooperatively), farm gate sales, community sup- ported agriculture and other forms of direct selling accounted in 2002 for 20% of organic retail sales in Germany, 16% in Australia, and 10% in Austria and Denmark. Similarly, direct sales accounted for 10% of all organic food sales in the UK in 2002-03 (Soil Association 2003). Even in the USA, where only 3% of organic retail sales were direct in 2001-02, thousands of farms are now involved in community supported agriculture programs that encourage tens of thou- sands of consumers to subscribe for a weekly share of the season's harvest, whatever it may bring. In contrast with the downward pressure that mainstream retailer involvement in organics is likely to place, in the longer term, on farm gate premiums, direct sales offer practical ways for farmers to capture a greater share of the consumer dollar. The emphasis placed by direct sales strategies on personal interaction, localization and seasonality makes them particularly well suited to smaller growers who struggle to offer the quantities and continuity of supply required by supermarkets. At the same time, the emphasis placed by direct sales on personal interaction, localization and seasonality promotes a sense of integrity and credibility in the organic sector. Altogether, this suggests that the relationship between large retailer involvement in organics and direct farmer to consumer sales is not always an antagonistic one. It is true that many of the people whom engage in direct sales, either as buyers or sellers, probably do so out of a belief that buying organic food at a chain store is somehow counter to the principles of organics. But it is also true: first, that many people who prefer to buy direct, when possible, still appreciate the convenience and accessibility of major retail chains when it is not; second, that many involved in direct sales as consumers came to an awareness of organics through the appearance of organic foods in conventional retailers; and third, that many small to medium size farms engage in both direct and indirect sales in order to spread their marketing options and risk.

The importance of large mainstream retailers and the visibility they lend to organics is underscored, review of organic food marketing in Europe and the USA. The most important feature of those supermarket chains that reported success in their marketing of organic products, in terms of profitability and corporate image, was the in-store visibility of organic products. This visibility was underscored by the range of organic products available, the availability of promotional and educational material within the store, and knowledgeable sales people. These chains took active steps to ensure that, wherever possible, every major product grouping included a high-quality organic option by appointing dedicated senior management personnel and by working with suppliers. The leading chains in organic sales had been involved in the sector since before demand for organic products mushroomed in the late 1990s (suggesting they were key players in this escalation of demand), had detailed environmental management policies and strategies that extended beyond the sale of organic foods to include matters such as energy conservation, and featured organics prominently in their advertising. Those chains less committed to organics tended to stock a limited range of long shelf-life dry goods. Almost all chains used their own organic certification labels to reduce confusion, although few collected data on who was buying their organic products or why. Consequently, there were few attempts to target the specific needs of different consumer groups. There was a general concentration on information related to health and food safety as well as, to a lesser extent, environment, taste and animal welfare.

Growing in terms of the Organic Food "Market"

The foregoing discussion has highlighted options for promoting continued expansion in the organic market. Consumer surveys from around the world consistently indicate both that most Western consumers are positively disposed towards organic foods and that the major barriers to their increased consumption of those foods are perceived to be their price and availability. The experience of major retailers in Europe is consistent with these claims. Those countries with comparatively more mainstream retailer involvement in the organic sector and/or lower retail premiums tend to have higher per capita levels of organic food consumption. At the same time, those retailers that report the most corporate benefit from organic sales are those that offer a comprehensive choice of easily identifiable organic product lines. Consumer research also highlights the importance of credibility and trust, matters that increasingly are addressed though the implementation of strictly regulated standards setting and inspection regimes.

We have no wish to imply here that all organic producers or processors need to do is get a certified product to market and it will sell itself. Although there is insufficient space here to discuss the issue fully, we suggest that one of the characteristics common to many growers, processors and retailers who have successfully expanded their sales is the considerable effort they have put into a supply chain organization. The key point is that resolving the supply side issues necessary to promote market expansion involves more than simply expanding production or putting more product lines on the shelf. For supermarkets that trade on convenience and choice, the focus of coordination is on securing supplies of consistent quality and quantity across a diverse product range. Strategies to achieve this range from supporting the conversion of conventional enterprises to organic production and processing, to encourage the pooling of produce from smaller farms into larger consignments, and importing supplies from elsewhere. Even farmers markets, community supported agriculture and box schemes that trade on seasonality, locality and community may be seen as methods to introduce convenience and choice to the realm of direct sales by making it easier for consumers to access a wider range of produce. Organizers of such schemes often supply recipes and other information to help consumers cope with the inconvenience of seasonal gluts and shortages and with unfamiliar produce.

As important as supply chain organization and standards regulation are to the organic market, we cannot overlook the questions of how organic foods are promoted and to whom. Organic foods are consumed in at least small quantities by enough Western consumers that there is little to differentiate between those who do eat organic food and those who do not. The values that underpin organic food and agriculture, environmental sustainability, food safety and nutrition, animal welfare and so on, are widely perceived as important. Nevertheless, the most committed consumers of organic food are those who place a high value on the perceived naturalness of food. These are more likely to be women than men, and more likely than not to take major responsibility for food shopping and preparation within their household. These groups do not need to be convinced of the merits of organic foods. Further, even though the demands placed on their time by responsibility for food provisioning lead this group to place a high value on convenience, this is the group most likely to actively seek out organic foods and to adapt their shopping behavior accordingly. In doing so, members of this group may challenge the idea that large supermarkets are as convenient as they are widely perceived to be and the notion that taking time to think about food is an inconvenience. In promoting organic foods to this group, the key attributes are naturalness and authenticity. They are uncomfortable with highly processed organic foods and long-distance transportation and would rather know where they can buy fresh local produce and minimally processed dry goods.

Most organic food consumers are not, of course, so committed. They do not change their shopping habits readily and will substitute conventional products for organic whenever the latter are not available or affordable at their regular shopping outlets. Nevertheless, increasing sales to this group may represent one of the easier growth paths for the organic industry. These buyers do not need to be convinced of the merits of organic food. They already purchase it. All that needs to be done is to convince them to purchase more of it. In order to promote organic products to this group the organic industry needs to address a range of issues including:

1. *Pricing:* retail price premiums remain considerably higher in most national markets than most consumers are willing to pay. Addressing this will remove a major barrier to organic food sales among less committed consumers.
2. *Visibility:* most food purchasing decisions are based on habit. Varying habitual patterns and developing new shopping habits is strongly influenced by the visibility of alternatives. Although introducing organic foods to major retail chains certainly has increased their visibility, the appearance and layout of product displays also is critical. The most influential area to display products is where buyers enter the retail outlet, at the ends of aisles, and/or at eye level. Even where this is not possible, it is important that displays identify organic products in some prominent manner.
3. *Labeling:* as well as reducing the visibility of organic foods, inconsistent and inadequate labeling reduces consumer confidence and trust in the integrity of organic claims. National and international harmonization or certified organic labels would be welcomed by most consumers if not by the certifying bodies that compete for farmers business.
4. *Availability:* the supply and quality of organic products must be consistent enough that buyers are not tempted to substitute them for conventional products. There may be circum- stances in which the organic industry is better served by a strategy of targeting a few key products than by attempting to provide a complete product range. Research has shown, for example, that a few fresh fruit and vegetables account for most of the expenditure, suggesting that the organic industry could have the greatest impact on its overall sales by targeting the top-selling items.

As stated above, there are relatively few Western consumers who state that they would not, under any circumstances, buy organic food. The barriers raised against organic consumption by members of this group typically focus, again, on pricing and availability. This group is clearly the least likely to bypass major retailers and make use of specialty food stores, home delivery services and so on. Given this, an obvious strategy to encourage members of this group to purchase organic foods is to sell them at the same price as conventional foods. At least one British supermarket chain has done this (selling some organic products at below cost) in order to increase the profile of its concern for the environment. However, it is likely also that many strictly conventional food consumers see little intrinsic value in the organic label and are unlikely to purchase any organic food that does not possess the same visible quality attributes as conventional competitors (e.g., visual appearance, smell) even if it is no more expensive. This suggests that it is necessary to promote to this group the benefits of organic products, a strategy that may encourage them to accept a higher price because they believe that the positive attributes of organic products such as health, taste and environmental sustainability increase its value. Higher in relation to those foods that were purchased more frequently, that were seen to be of higher quality and safety, and that were underwritten by trustworthy certification schemes. Taken together, these suggest that the best way to maximize organic market share will be to reduce retail premiums on more expensive value-added items while improving the quality and supply of less expensive, but more frequently purchased, items for which higher premiums can be maintained. However, organic foods often are represented, positioned and priced as premium gourmet brands, a strategy that may limit overall levels of demand.

Third, changes in the retail structure of the organic industry are leading to major changes in the availability and cost of organic foods. While health food and specialist organic stores pioneered organic retailing, mainstream supermarkets now dominate the sector. This shift has contributed to lower retail price premiums and higher growth rates in the total market share of organic foods. But it appears also, in places, to have contributed to downward pressure on the prices received by farmers. Pressure to increase efficiency and economies of scale within the production sector in ways that may be inconsistent with organic values and principles. Supermarket dominance is also likely to promote increased reliance on imports of organic food from lower cost production centers in Africa and South Amer ica. Fourth, in light of evidence that retail price premiums in general are substantially higher than those premiums received by farmers, it seems less than surprising that alternative marketing strategies based on direct farmer to consumer sales are developing alongside supermarket interest in organics. These promise to lower the gap between farm gate and retail premiums and suggest a range of possibilities for the resolution of supply problems and expansion of demand.

In sum, while there are plenty of signs that demand for organic foods will not support indefinite growth given existing retail price premiums, there also is evidence that innovations in distribution and retailing, coupled with lower retail price premiums, may provide the basis for continued expansion well beyond existing levels of organic food consumption.

10

Organic Farming's Social Responsibility
Education, Cooperation, and Regulation

Social justice has not been an integral part of the construction of the organic movement. Yet, today we are witnessing the emergence of new movements or perhaps a rejuvenation of existing movements, which seek to promote fair trade and new codes of conduct. The introduction of new concepts such as chain transparency, food chain awareness, food miles, just food and sweat shops, and the appearance of bestseller books like Fast Food Nation, suggest an increased public concern about food that transcends health, taste and price, to include the social costs of food production.

This Chapter provides an overview of social responsibility within the organic sector, referring to different conceptualizations and discussions underway. Against this backdrop, the focus on social justice issues, as embraced by leading organizations in the field is discussed (see social responsibility in the organic context), and a three-pronged approach for addressing social responsibility is presented and advocated (see A framework for approaching social responsibility in Organic Farming). Regulation, collaboration and learning are argued as three necessary and (potentially) mutually reinforcing coordination mechanisms for fostering enhanced social responsibility throughout the organic sector and furthering the social agenda. Regulation and collaboration are explored, with several examples provided to illustrate current manifestations of how these mechanisms work. These are the more developed and active forces within the sector. A learning approach is articulated in more depth (see Learning and social responsibility), with nuances and challenges considered.

Because learning is a less-developed approach, we make a case for learning to be embraced more fully. The argument draws on learning theory and applies this theory to the reality of the challenges facing the sector with respect to social responsibility. Two types of learning are discussed: instrumental and emancipator learning. Instrumental learning is learning for social responsibility, whereas emancipator learning is learning towards social responsibility. Discussion on emancipator learning describes the spectrum from big brother to grassroots learning and explores the role of conflict in this. Further, different levels of learning from inter stakeholder to intra stakeholder and from micro to meso and macro are all touched upon. A final section reasserts the value of the three-pronged approach advocated and offers an example, internal control systems (ICS) for group certification in developing countries, to illustrate how regulation, collaboration and learning can overlap and reinforce one another to the benefit of all. This Chapter concludes with a reiteration of our argument that values the utilization of all three coordination mechanisms, and places the discussion back into the larger agricultural and societal contexts.

Organically-based Social Responsibility

It is no surprise that the technological revolution that swept through agriculture in the 1950s fueled questions by some of the pioneers of Organic Farming. After all, they were concerned with the health of the soil, plants, animals and people, and were committed to maintaining farms that were small in scale, self-sufficient and accepting of the natural order. The commitment of such individuals and discreet organizations transformed Organic Farming from a rather loose concept into a social movement with associated farming methods in the last quarter of the 20th century. This movement intentionally placed itself as an alternative to mainstream agriculture. The consolidation of fringe movements and subsequent evolution of these movements into a more coherent whole, which took place in the second half of the 20th century, was very much the result of a two-way process of societal demands shaping Organic Farming, and the organic movement gradually influencing some sections of society. Over time, the philosophical roots of the movement have been translated into standards to guide systems of production and, with the rapid expansion of markets and the appeal of high prices, the organic movement has emerged globally as a recognizable sector within agriculture. Producers and producer organizations, certifying bodies, processors and traders, service providers, various organic interest groups and a plethora of other players now constitute this sector.

The international growth and development of the organic sector can be seen as positive in terms of its contribution to sustainable agriculture and the triple goals of social responsibility, economic viability and environmental integrity. However, the global nature of this expansion and the likely absorption, appropriation and concentration of Organic Farming by conventional agriculture and its market structures, the so-called conventionalization of Organic Farming, and its consequences in terms of the original ideals of the organic movement, have been topics of recent intense discussion. The most rapid growth of organic farming has taken place in Europe. Growth as a break through, whereby what was an obscure type of farming in the beginning of the 1980s had become an institutionalized form of production with its own dynamics in the market place by the year 2000 in some countries. The rising influence of the environmental movement over the same period leading to greater public interest in environmental matters, coupled with the prominence of a series of food safety concerns, are attributed to heightening consumer interest in organic products. The formal political recognition of the organic movement and ongoing policy formulations at the European Union (EU) level has complemented this growth process.

In the consumers' minds, personal well-being, taste, environmental values of a general nature and animal welfare concerns have been the more important motivations for buying organic food as opposed to an adherence to organic principles. Social values and social justice criteria have not been expressed as distinct driving forces for consumer behavior towards Organic Farming in Europe. In a country like Denmark, this has the highest consumption of organic products per capita in the world; a prerequisite for such high consumption is seen as a trustworthy and recognizable organic label, which guarantees that organic production rules have been observed. Public good attributes, such as improved animal welfare and environmental protection, were seen by most respondents as important features of organic production, even higher than private good attributes such as health. In the United States of America (USA), Social and environmental relationships are not separable in practice, but went on to say that there was a disturbing invisibility of important social issues, such as labour relations and land tenure, within conventional agriculture as well as within its organic counterpart. Social responsibility considerations may have been integral to the visions of pioneers in the organic movement, both producers and consumers, but these have been neither well-articulated nor codified. In 2002, the social justice chapter of the organic standards for standards.

Institutionalization of 'Organic' Values

Since the 1920s in Europe, a distillation of organic values, ideology and principles of the social movement have developed into standards, rules, regulations and eventually national legislations. The development of the sector from locally entrenched fringe movements to a more widely accepted alternative to conventional agriculture is a triumph, but not without compromises. The translation of organic movement values into measurable indicators for standard verification and legal requirements has allowed industrial agriculture to enter into organic production. As a result, some argue, value-laden private sector standards are becoming input-based recipes and prescriptions. There has been much debate and controversy over what has been termed the institutionalization, conventionalization, commoditization, scaling up or mainstreaming of Organic Farming. Concerns abound that the widespread institutionalization of Organic Farming is leading to an erosion and even loss of the movement's core values.

Figure 10.1 illustrates the relationship of Organic Farming within the agricultural sector and to society at large. Concepts of Organic Farming are socially constructed and circulated in a dynamic way that continually mirrors and reshapes the contexts in which the production and consumption of organic produce occurs. This is also true of the social aspects of Organic Farming. The triangle housed within the organic sector (Figure 10.1) represents the social responsibility aspects of the movement; the three sides denote the three approaches we put forward as complementary (see a framework for approaching social responsibility in Organic Farming): learning, collaboration and regulation. The relationship to broader society and the agricultural sector is relevant as different perspectives of social responsibility within the Organic Farming movement are explored. Despite the inception of Organic Farming as a counter movement to the modernizing processes of agriculture in general, recent concerns have been expressed in Europe and the USA about Organic Farming beginning to resemble and being incorporated into conventional agriculture. In a more global context, particularly that of poorer farmers in developing countries, raise the question as to whether it was important for Organic Farming to be retained as an opposition to modernizing processes.

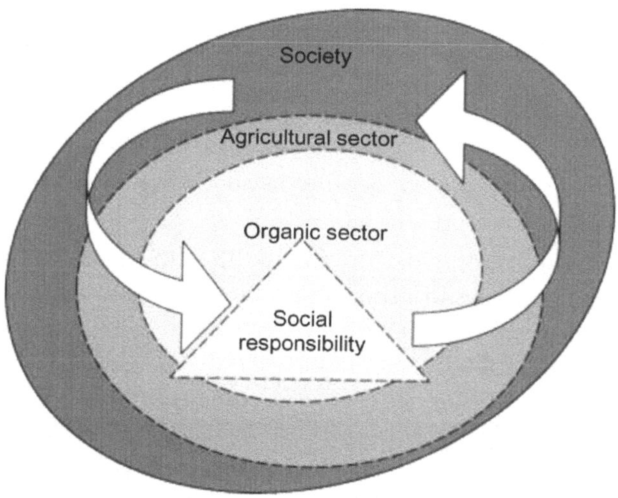

Fig. 10.1: Positioning social responsibility in the organic sector. The arrows indicate the interconnectedness of the various contextual levels

As part of setting its directions for research and development of the organic sector, the Danish Research Centre for Organic Farming (DARCOF) identified three basic principles for action: the cyclical principle, the precautionary principle and the nearness principle. The nearness principle implied that social aspects of food production can be improved by nearness between producer and consumer. Applied at the local level, as indeed had been taking place when the organic industry was small and agriculture very uniform in a country like Denmark, production, processing and marketing systems were proximate and transparent, and social justice concerns were hardly expressed. If the nearness principle is extended to the wider organic sector at national and global levels, it seems sensible to argue that emotional nearness would substitute for physical or geographical nearness when it comes to social responsibility concerns.

Localization as an alternative to the globalizing process of agriculture and as a way of re- establishing the proximity between producer and consumer has been attempted in Organic Farming and through several other related initiatives, such as farmers markets and community supported agriculture. On the basis of studying 37 local agrifood initiatives in California, which as a State has a long history of such initiatives, concluded that social justice may be difficult to construct at a local scale. Despite the wide recognition of the issue of social justice of labour as problematic in California, these researchers found that social justice concerns of the agrifood initiatives they studied were more to do with food access and support for small farmers than with justice for farm workers in terms of recognition, wage and job security. Questions related to the position of seasonal workers in organic farms in Europe have not been addressed through research either. Paucity of information on social problems in organic farming and the lack of in-depth studies on working conditions. A Danish study on changes in the working environment for organic farmers revealed that not only the physical workload tends to increase on organic farms, but also the amount of psychosocial stress. The latter is mainly the result of the additional administrative workload and dealing with changes in market conditions, farm size and employee relations.

On the consumption side too, inequalities in access to food seem to be brought out by organic and other alternate food systems. The higher prices for organic food compared to conventional food make organic food beyond the reach of poor people. It was the industrialized food systems that reduced income-related class differences in food consumption and made food more accessible to all. Ironically, the ways in which the emerging alternate food systems are structured threaten to reverse or break down this leveling, commenting on the new rural development paradigm of Western Europe and the new modes of food provision within it such as short food supply chains, speaks of these initiatives as overlooking social justice issues. In the long run they will favor a new multitier food system differentiated by income and class, and quality and nutritious food as the missing guests at the table. Ironically perhaps, there are now low-budget supermarkets such as the ALDI Group that offer very affordable organic products for a specific niche of price aware, light green, soft bio consumers. However, these same low-budget supermarkets have a poor track record when it comes to workers' rights and providing adequate worker compensation.

The driving principles of low-budget supermarkets are obviously different from those of the certifying bodies of organically grown products. If there is a demand for a product, they will find a supplier. This leads to dilemmas for the organic farming sector. Berlin-based Märkisches Landbrot, one of Germany.s largest organic bread and muesli producers, typifies a perhaps overly principled stance. The company will not offer discounts to supermarket chains that are looking for organic suppliers. Märkisches Landbrot's marketing director Sabine Jansen says that to offer cuts to the big chains would do in the natural food stores'. For the same reason, Demeter is unwilling to supply goods it certifies to nationwide discount chains like Plus or ALDI. The result is that the organic products that fill Germany's bargain grocery chains and upscale bio supermarkets increasingly come from somewhere else.

A question arising from the social justice concerns referred to above in connection with contemporary Organic Farming sectors in Europe and the USA would be the degree to which some of these inequalities will be reproduced in developing countries as their organic sectors begin to be shaped by the consumption patterns and institutional structures of the developed countries. Rising world organic trade and the expanding agrifood networks between developing and developed countries. The rigorous controls in place and the developed countries market demands for continuous supply of quality goods no doubt leave the small producer in developing countries with few alternatives. The integration of these small-scale peasant producers into the Fair-Trade network, and sees possibilities at the global level for greater convergence between the organic and the Fair-Trade movements. In addition to Fair Trade links between producers in the developing world and consumers in the developed world, recent cases built on Fair Trade principles that operate exclusively between developing countries or exclusively between developed countries.

Initiatives defining and addressing Social Justice in the Organic Sector

Many in the organic sector are disappointed with what may be seen as the reduction of the movement's core values into technical packages of practices and prescriptions as to how to farm without chemicals. Through the Victoria meeting, a social justice survey was developed and disbursed with the intention of mapping the landscape of social justice within sustainable agriculture, seek out opportunities for collaboration and strengthen social justice in sustainable agriculture. Some of the themes addressed in the survey were:

- General information social accountability;
- Social standards and standard setting;
- Using standards and verification of standards;
- Formal and informal verification systems;
- Use of Internal Control Systems;
- Formal certification systems;
- Social certification and organic certification systems;
- Trading relationships;
- Trade unions and workers associations;
- Capacity building, market development and advocacy;
- Research; and
- Building a social justice community.

While the survey was ambitious, the response was disappointing with only 38 respondents. The Bangkok meeting revealed several gaps in organic and sustainable agriculture standards, among which were the following:

- Mechanisms needed for implementation of social justice standards
- No means for farmer price-setting outside market forces
- No fair-trade mechanisms in developed countries
- Limited capacity among farmers to set standards
- No bottom-up strategies in place for evolving standards
- No definition of child labour within the reality of the farmers
- Lack of food security mechanisms for small farmers
- Lack of standards for buyers-community/rights and responsibilities/pricing
- Lack of recognition for good practices among family/traditional farmers
- Assessing whether standards undercut traditional market models
- No discussion of wild harvest products
- No development element
- Undocumented workers must be more comprehensively addressed
- Better training for social justice needed (e.g., auditor training for social skills)
- No reference made to price setting and subcontractor/operator relationships

- Land tenure of indigenous people needs to be addressed
- Lack of technical assistance for compliance to producers
- IBS do not address farmer's contracts with buyers
- Standards address gender discrimination, but are not explicit about cultural values
- Uncertified organic or de facto' organic is not addressed in standards
- Regulations needed for the negotiation of fair contracts, and
- Farmers need access to government programs.

These issues capture the constellation of concerns that are part of the social agenda of the organic movement. Different organizations and individuals are working on different aspects of these concerns and issues in many different ways. The Rural Advancement Foundation International (RAFI) and IFOAM are among them, and their efforts are outlined in the following paragraphs.

RAFI has developed a comprehensive set of social stewardship standards, with the intent of reincorporating social justice in sustainable agriculture and to address the gaps of other standards and growing concerns of many in the organic sector. The standards are divided into the following sections: farmer rights and buyer responsibilities; buyer rights and farmer responsibilities; indigenous people's rights; farm workers' rights; interns and apprentices. It is uncertain as to whether another label will emerge from these efforts or whether instead the promoters will continue to work with other standard setting organizations to improve the standards that are already in place. IFOAM, as the membership-based international voice of the organic sector, has participated in the social justice meetings both in Victoria and in Bangkok, as well as supporting the development and implementation of the survey described above. In addition, IFOAM has undertaken several key initiatives to address the social agenda within the organic sector, including the development of a code of conduct for traders, participation in the Social Account- ability in Sustainable Agriculture (SASA) Project, coordination of three years of meetings to address smallholder certification, ongoing communication and collaboration with partners in fair trade, the development of the Organic Traders Charter and so on.

The SASA Project definition of social responsibility issues areas are well recognized as key social justice concerns internationally (SASA 2005). Key issues addressed in the project include:

- General information social accountability;
- Freedom of association and right to collective bargaining;
- Working hours;
- Seasonal workers, contracts and undocumented workers;

- Child labour;
- Health and safety;
- Wages/compensation;
- Discrimination;
- Basic treatment and disciplinary practices; and
- Forced labour.

Although the breadth of social issues and the many conceptualizations of social responsibility in the organic world are outlined earlier in this chapter, the SASA issues listed above are the main components of social responsibility considered further.

Table 10.1: Three perspectives encountered in different social science discourses

Perspective	Government-driven	Market-driven	Community-driven
Rural policy practice (Rob Schrauwen pers. comm., 2005)	Regulating	Compensating	Stimulating
Rationalities (Habermas 1984)	Instrumental	Strategic	Communicative
Bases for individual behavior change (Kelman 1969)	Compliance	Identification	Internalization
Preferred ways of arranging human affairs (Hood 1998[A])	Hierarchy	Individualism	Egalitarianism
Organizational coordination mechanisms (Powell 1994)	Hierarchy	Market	Network
Causes of wealth of nations (Bowles and Gintis 2002)	Resources (such as power or access to natural resources)	Invisible hand of market forces	Social capital, trust, community
Institutionalization (Giddens 1984)	Domination, legitimation	Liberalization, persuasion	Signification, communication

A Framework for approaching Social Responsibility in Organic Farming

The three ways of getting things done are modified to illustrate the current and envisioned options for social responsibility within Organic Farming. Rolling distinguishes several change perspectives that arise in the social sciences, drawing out similarities in the approaches. The first perspective emphasizes hierarchy and (government) authority, and employs an instrumental approach that includes compliance to regulations. The second perspective puts emphasis on market forces, economic incentives and compensation, and employs a rational choice approach with strategic underpinnings. The third perspective relies more on learning, stimulating change, internalizing concepts and building social capital and action among equals.

The three dimensions discussed in this Chapter are represented in figure 10.2: learning, collaboration and regulation. The three dimensions are complementary but distinctive, relying on diverse coordination mechanisms, motives and motivations. Learning is a stimulating approach and regulation provides a framework, while collaboration entails communication with other stakeholders (organizations), recognizing available expertise and building relationships that benefit the many people engaged. In creating and revising regulations, and when collaborating, learning will inevitably take place among those invited or encouraged to participate. Although the nature and kind of learning taking place can vary, not all stakeholders are always invited. The three approaches are most fruitful when working together in a dynamic way with differences in emphasis depending on the context in which social responsibility is developed. Regulation is seen both as an outcome and an input for (further) learning, while regulation, or rather the process leading to regulation, is also seen as a learning process.

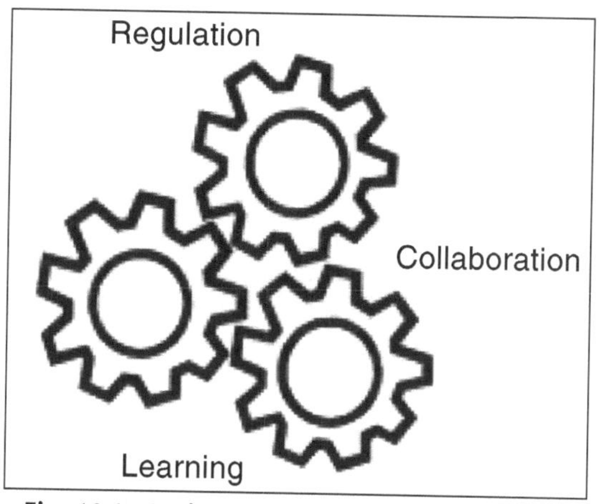

Fig. 10.2: An integrative approach towards fostering social responsibility in the organic sector

Regulation

Regulation is an instrumental approach depending on a hierarchy of standards, legal regulations, accreditation systems and international guidelines. This approach to social responsibility depends on compliance to social standards set by stakeholders, some within the organic sector, and others beyond it such as national laws and the International Labor Organization (ILO). Compliance too widely (internationally) recognized social standards is key to the success of this coordination mechanism. The strength of a regulatory approach is that it is binding, more universal (within a given frame of reference, e.g., a nation), is third party verified, and it causes social responsibility and social justice to become entrenched in the system. Regulation is already a reality in the organic sector and as such, the regulatory framework is outlined in the following subsection. The use of regulations and standards to set parameters and delimit the playing field is indeed valuable.

Current Regulatory Framework

As the organic sector developed from multistranded, diverse and geographically and culturally distinct movements internationally, in order to assure integrity and retain value for the practices undertaken, ideals have formalized, being articulated as principles, standards, private certification and later as regulations/legislation and international guidelines or standards. The Demeter label, first used in 1924, marked the beginning of this non-linear development process. There are about 360 private certification bodies internationally, many with their own standards for people certifying to US, European and Japanese regulations to allow access to those markets. As of December 2001, 56 countries were at different stages of organic regulatory development, 32 with fully implemented regulations. The IBS were first published in 1980 and have been revised and voted on every two years (every three years as of 2002) by IFOAM members. IFOAM is working to harmonies both private standards, encouraging mutual recognition among accredited certification bodies (ACBs), and national regulations, to facilitate international trade in organic products. In addition, a joint FAO/WHO intergovernmental body established in 1972, adopted organic guidelines in 1999 and livestock standards in 2001. These private and public definitions of organic set the parameters that govern producers and processors.

Compared to production aspects, the human side of organic farming has had less attention in regulatory developments. While social responsibility was a central tenet to many organic movements, it has been sidelined somewhat as the movements consolidated into a rapidly growing and diversifying sector. That said, IBS include a short chapter (8) on social justice. These standards for standards apply to all ACBs accredited with IFOAM's International Organic Accreditation Service (IOAS). ACBs accredited by the IOAS will be inspected during their verification visit as to how they are implementing these social standards. The following excerpt reflects the IBS chapter on social justice (IFOAM 2002). These social standards are basic compared to the more comprehensive standards of other social certification and standard setting organizations such as Fair-trade Labeling Organizations International (FLO), Social Accountability

International (SAI), Sustainable Agriculture Network of the Rainforest Alliance (SAN), and the ILO. All IOAS-accredited certification bodies are expected to develop their own social standards based on these broad IFOAM standards for standards. Some private organic standards (IOAS accredited or not) already incorporate social standards into their certification. Often producer associations or cooperatives in developing countries are subject to social certification via fair trade in addition to their organic certification.

Collaboration

Collaboration as discussed here refers to the engagement of multiple stakeholders with each other in the organic sector covering the entire supply chain to better address social justice issues. The middle column tends to refer to economic and market approaches. Collaboration is included because in addition to being an approach to (social) learning, collaboration is also very strategic and has potential cost-cutting and other economic benefits. In this way collaboration can be regarded as a rational choice approach towards social responsibility. At least two kinds of collaboration can be distinguished: collaboration among social and environmental certification organizations, and collaboration throughout the supply chain. The former refers to those in the organic sector working with experts in social certification. In this example of collaboration, the intent is not to develop comprehensive social criteria, but rather that the organic sector works with other certification systems that specialize in social issues. The second kind of collaboration refers to coordination throughout the supply chain and as such, is closely linked to the first. In order to ensure that social standards are upheld along the supply chain, coordination is required. An example is the complex organic cotton supply chain that involves among others: producers, ginners, spinners, dyers, exporters, cut/ make/trim operators, designers and retailers. Each step holds different ecological and social challenges and demands different skills to adequately address the risks.

Collaboration with organizations more experienced in social certification is a wise and vibrant endeavor for IFOAM. IFOAM, via the International Social and Environmental Accreditation and Labeling (ISEAL) Alliance, has a platform to generate and learn from such synergistic relationships. However, while representing the organic sector, IFOAM is not the only participant: certification bodies, researchers, producers, consumers and other stakeholders are also challenged to collaborate and make links between ecological justice (which the organic sector does well and has experience in) and social justice.

Learning and Social Responsibility

There is neither a single outlook as to what social responsibility entails nor a clear pathway marking how it can be achieved in the organic sector.

As such, there are different ways to respond to the challenge. One response, a regulatory approach, would be to look for the core meaning and components of social responsibility in Organic Farming that are agreed upon (by some) and prescribed (to others) on a global scale. Discussion then would focus on questions to decide who should determine this core, how an agreed upon core can then be best transferred/communicated so that it is taken up by others and how compliance to standards and guidelines for practice derived from this core can be monitored throughout the entire supply chain. Through negotiation and consensus seeking, experts and (other) stakeholders can arrive at mutually suitable methods and processes for operational sating social responsibility that can be implemented and enforced on a global scale. Indeed, this has been the approach in developing IFOAM Basic Standards, as well as the broadly recognized social issues that are taken up by leading social certification organizations (see Initiatives defining and addressing social justice in the organic sector). This response is widely practiced.

Table 10.2: Emancipator versus instrumental learning

	Instrumental	Emancipatory
Nature of knowledge	Static, finished (scientifically) agreed upon, universal	Dynamic, under construction, contested, contextual
Process of learning	Transmission, instruction	Transformation, co-creation
Role of learner	Consumer/user of knowledge	Producer/user of knowledge
Role of initiator	Instructor	Facilitator
Origin of learning goals	Pre- and expert determined fromthe outside	Self-determined from within
Object of learning	Tailor made for specific target groups	Multiple stakeholders in an integrative setting

Another response to the many perspectives on social responsibility regarding Organic Farming is to acknowledge that working towards social responsibility involves questions regarding values, ethics, justice and equity. Pathways towards social responsibility are unlikely to develop without friction, controversy and conflict. This view emphasizes that we live in pluralistic societies, characterized by multiple participants with divergent interests, values, perspectives and constructions of reality. Furthermore, working towards social responsibility demands more than universally applicable recipes for social responsibility. Governments cannot rely on the exclusive use of legislation to enforce social responsibility.

Most countries have basic laws regarding wages, working hours and child labour among others, but this does not necessarily mean they are adhered to. Standard-setting bodies cannot depend only on economic instruments (e.g., premium prices) to stimulate the whole supply chain to act in a socially responsible manner. Genuine social responsibility develops from within and is anchored in values and a deep conviction of what is considered right and just within a given subculture. The question then becomes more complex as to how can internationally recognized standards integrate with divergent local values (and vice versa). International standards, although they are more flexible than laws, are less dynamic than local standards.

Innovation and change in land use, production; processing and consumption can be seen as a result of the learning of individuals and groups active in the agri food chain. Likewise, learning can generate innovation, change and development of social responsibility in (organic) agriculture. There are many ways to conceptualize the relationship between learning and social responsibility, but for this Chapter we address two possibilities: instrumental learning and emancipator or social learning. The main differences between instrumental and emancipator learning lie in: the nature of knowledge, the process of learning, the role of the various stakeholders in learning, the origin of the goals that are being pursued and the object of learning (Table 10.2). When exploring social responsibility by using this dichotomy, again two pictures emerge: learning for social responsibility (instrumental) and learning towards social responsibility (emancipator).

Learning for Social Responsibility

The for suggests that there is something known and agreed upon when it comes to social responsibility in organic farming. Indeed, there are legal and regulatory frameworks within which Organic Farming operates. As well as IFOAM's social standards and private certification schemes, the Charter for Organic Trade and the ILO guidelines among others address elements of social responsibility. Experts, policy-makers and special interest groups have influenced these frameworks, schemes and codes, and seek to get producers, processors and traders to adopt them. Through information, extension, persuasion, rewards and punishment, relevant participants can be moved to comply with the standards, norms and guidelines that have been set. This type of learning can be categorized as instrumental. It relies heavily on the regulations and standards already in place.

Learning towards Social Responsibility

Towards suggests that while there is a constellation of ideas as to what social responsibility entails, the lack of consensus about the implications of an exact meaning in variable contexts prevents global prescriptions. Although there is certainly agreement about basic components of social responsibility and social justice, more nuanced understandings and applications need to be constructed contextually in an integrative process involving all stakeholders. Some would argue that consensus about an ill-defined issue such as social responsibility is undesirable from a radical democracy perspective. Radical democracy offers a way of thinking about difference, as opposed to seeking consensus. Democracy, from this perspective, depends on differences, dissonance, conflict and antagonism so that deliberation is radically indeterminate. In this view, the conflicts that emerge in the exploration of social responsibility are prerequisites rather than barriers to reaching more sustainable solutions. We should acknowledge and valorize the diversity of ways in which the democratic game can be played, instead of trying to reduce this diversity to a uniform model of citizenship. This would mean fostering a plurality of forms of being a democratic citizen and creating the institutions that would make it possible to follow the democratic rules in a plurality of ways. Social responsibility can be seen as a process that negotiates language meanings and cultural understandings between equal parties with equal access to the negotiating table. Where such equality is lacking, measures need to be taken to overcome inequalities to create more optimal conditions for social learning. As such, social responsibility can be viewed both as an evolving product and an engaging process. Social responsibility as a social learning process deserves more attention as do the more accepted concepts of social responsibility as an expert predetermined transferable product. If the premise is accepted that there is neither a single outlook on what social responsibility entails nor a process to achieve it, then one might also accept that determining the meaning of social responsibility is a process that involves all kinds of stakeholders in many contexts, people who may not agree with one another. There are different levels of self-determination, responsibility, power and autonomy people can exercise while engaged in such issues or disputes. In dealing with conflicts about how to organize, consume and produce in socially responsible ways throughout the organic supply chain, learning does not take place in a vacuum but rather in rich social contexts with innumerable vantage points, interests, values, power positions, beliefs, existential needs and inequities. The amount of flexibility that individuals have for making their own choices, developing possibilities to act and for taking responsibility for their thoughts and actions, varies.

A continuum can be used to indicate the different levels of self-determination, self-responsibility and autonomy people can exercise within social responsibility-oriented learning processes that are directly related to the individual's room to man oeuvre. Another continuum can be distinguished, one expressing the degree of openness and the extent to which outcomes and processes are prescribed to citizens. The resulting force field (Figure 10.3) shows four different conceptualizations of social responsibility and the role of citizens in moving towards social responsibility. The upper left quadrant of the force field can be named big brother or instrumental social responsibility: social responsibility that has been authoritatively determined and defined by experts and prescribed to relatively obedient and passive citizens.

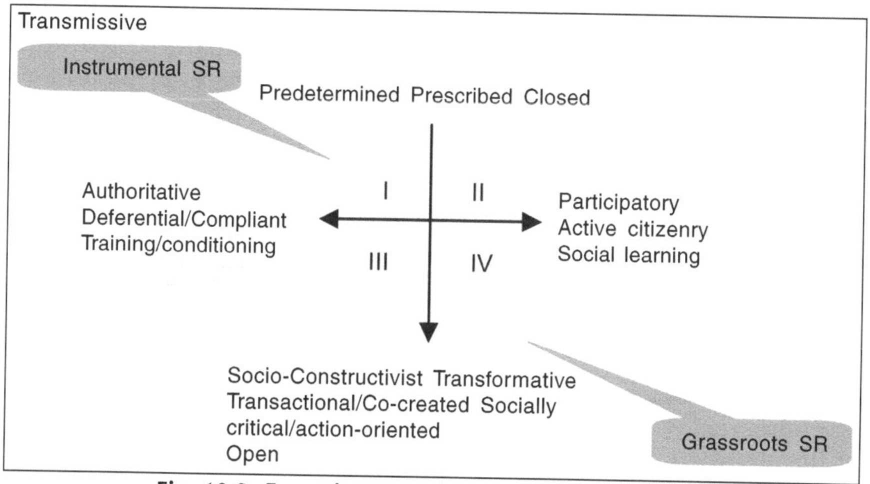

Fig. 10.3: Emancipator and instrumental perspectives on fostering social responsibility in the organic sector

The lower right quadrant represents grassroots social responsibility, characterized by the active and critical involvement of competent and action-oriented citizens with high levels of self-determination in finding pathways towards social responsibility as agreed upon by all, and high levels of empowerment and self-actualization. Whether the resulting society is sustainable or would be viewed as socially responsible from the outside is questionable, but people might be happier with their situation (as compared to the first scenario), at least for a while.

The other two cells represent forms of social responsibility characterized by a limited openness and/or involvement. This can be seen, for instance, when groups are encouraged to actively participate in the implementation of a guideline for social responsibility that has been predetermined without their involvement. Limited or even false participation is often the result. Figure 10.3 as presented here is perhaps normative with a bias towards openness and participatory democracy. The bias is somewhat contrived when considering, for example, the role of expertise. In the upper left quadrant there is a tendency to downplay ordinary peoples voices and to stress expert knowledge (the experts know what's right, the others know nothing), while in the lower right quadrant there is a tendency to downplay expert knowledge and to stress the ordinary peoples' voices (leave it to the people and their own local knowledge, without outside interference). A more fruitful approach may be to consider each quadrant as a potential starting point towards social responsibility and to recognize that all quadrants have something valuable to offer. Conflict in social responsibility.

In considering the levels of self-determination illustrated in table 10.2 different types of potential conflicts can be envisioned to emerge as stakeholders endeavor to move towards social responsibility. The two extremes are the big brother instrumental approach and the grassroots emancipator approach towards social responsibility and citizen involvement. In big brother social responsibility, conflict is avoided or kept under the table internally, by using objective, rationally determined standards or norms. An abundance of social standards, guidelines and regulations to achieve this can be cited (e.g., ILO, Fair Trade, SA8000, and IFOAM IBS chapter 8, national labour laws). After determination of the norms and standards, a range of instruments are used for implementation or enforcement: laws, legislation, regulation, reward and punishment schemes, (mandatory) training and instruction or financial incentives. These instruments leave people little choice but to adapt the expected behavior, if they aspire. Conflict might emerge internally when setting the standards and deciding on how to implement them, but once agreed upon, any doubts about the standards are masked as much as possible. This gives producers and consumers confidence in the reliability of the scheme. When rules are broken, standards are either not met or are met superficially, or critiqued by certain groups, conflict does emerge eventually among those affected by the rules, regulations and standards.

In grassroots social responsibility different perspectives, kinds of expertise, values and interests converge in a consensus-seeking process in which conflict is inevitable. Cultivating conflict and using conflict as a force for conceptual change and creative problem solving is a prerequisite for arriving at solutions that people can identify with and act upon. Ownership of solutions is thus engendered. Social responsibility in this approach is rooted in local contexts taking on many forms and shapes as contexts change geographically and over time. Social responsibility develops on the edges of carefully facilitated and sometimes mediated encounters between different interests, values and world views. In an ideal social learning process, all participants involved jointly arrive at a temporary vision of social responsibility which they share and identify with. In the last analysis, social responsibility in this context comes more from within, rather than from outside. There may not be consensus about everything but there might be a renewed sense of community and interdependency, and even respect for differences or respectful dissension and radical democracy. This quadrant has its flip side. There will always be differences in knowledge bases, access to resources, including networks and therefore inequities which lead to conflict and even demand more instrumental guidance and/or outside expertise, if only to help overcome these inequities. Likewise, big brother learning has its positive side where it can make transparent and transferable the values/standards necessary for certification.

Inter and intra Stakeholder Learning

Bettering the condition of our farm workers shouldn't fall solely on the farmer', says Muller, a 25-year veteran of organic farming. Everyone in the food chain needs to adopt a sense of fairness and responsibility for the well-being of farm laborers. It needs to be a partnership through the whole agriculture system, with wholesalers and consumers paying fair prices that then assure that farm workers are adequately compensated in an equitable way. The equation of greater social responsibility needs to be integrated though the whole food system.

When viewing the role of learning with respect to social responsibility, the levels of learning need to be identified. Three such levels can be distinguished: micro, meso and macro. The individual learner (i.e., the farmer) who learns from experience or from others, particularly from other farmers, is placed at the micro level. Micro level learning can be both informal and more formally organized (e.g., in farmer field-schools, farmer groups for co- learning, farmer study groups). At the meso level are communities of learners at local and regional levels or at a particular point in the supply chain (i.e., learning within cooperatives, farmer associations, trade unions, processing facilities)

Learning at the meso level requires that multiple stakeholders meet somewhere or somehow. Here too, informal and more formally organized learning can take place. Examples of more informal social learning at the meso level include farmers markets where consumers and producers interact and learn from the interaction, or microfinancing schemes where policy makers, intermediate institutions and poor farmers interact around loan schemes or where standard-setting organizations work jointly on a project to address shared issues. The meso level may also be relevant within a particular circle (i.e., producer associations, trader forums, unions, consumer associations, platforms of ACBs, standard-setting organizations) and can become a first step towards more systemic learning (e.g., subsystem learning). The SASA project is an example of systemic learning.

At the macro level there is systemic learning that takes place within an entire sector, involving all stakeholders in the supply chain and certification hierarchy. A key assumption underlying this type of learning is that the whole system is more than the sum of the participants and their relationships making up the system, and that systemic innovation can take place only with the full participation of all contributing to the system.

When looking at social responsibility issues, many require a systems approach to arrive at sustainable solutions. Macro level learning is complex and hard to organize, and examples of multilevel, systemic, chain-oriented learning (MLSCL) are rare, certainly within the context of organic farming.

Most multinational companies such as McDonalds, Unilever and Nike, considered to be operating successfully at a global scale, owe their success to creating tightly knit chains that respond quickly to changes in the global economy. Although the added value of these chain-oriented companies tends to be limited to economic value and the distribution of the added value can hardly be called equitable, these chains seem to be very effective. There are some examples of MLSCL-learning in watershed or catchment management schemes that involve multistakeholder social learning. Examples of MLSCL within the organic farming context that seek to create added value not only economically, but also ecologically and socially within the entire chain, tend to be regionally oriented, although there are also some transnational examples. The development of social responsibility can be seen as a dialectic between intrinsic (personal values) and extrinsic motivations (external pressures). This dialectic occurs at all three levels.

A Three-pronged approach to Social Responsibility

Although so far this Chapter has focused predominantly on the learning approach towards social responsibility, we advocate a mix of coordination mechanisms to stimulate, regulate and provide incentives for social responsibility throughout the supply chain. By providing stimulating learning and cooperation among organizations that promote social responsibility, and by solidifying that what is learnt in agreed-upon and broadly supported rules and regulation, a three-pronged approach serves the sector well. Although learning is a soft tool that allows room for errors, local variation, differing conceptualizations and the generation of a sense of ownership and internalization of values, regulations delimit the playing field, also for those unable to participate in the learning and collaboration processes described in this illustrates how the three coordination mechanisms can work together providing a basis for change and the development of stronger social responsibility in the organic sector.

The ICS combine the three coordination mechanisms put forth in this Chapter as potential approaches towards social responsibility, at the farmer level, or more specifically, at the development country small farmer cooperative level. By drawing in and developing the learning element of internal control systems, producer organization ICSs can become learning organizations with systemic learning as a key component of their operational process. The ICS is an example of a meeting point between grassroots and big brother learning. At the grassroots level, farmers learn about technical aspects of organic production, learn to organize themselves (in a cooperative in some cases) and provide documentation of their farming practices. In addition, a structure is in place through which community initiatives can be pursued (e.g., a Thai rice cooperative wherein self-sufficiency objectives were met via the ICS structure). At the big brother level, the rules are set by the certification body; however, the producer organization must adapt and internalize the standards by creating a manual that expresses the group's goals, localizing them to reflect local challenges and risks.

Regulations and standards are necessary parameters, especially when they emerge from a multistakeholder social learning process, and when they are not carved in stone but seen as flexible outcomes in need of constant revision. They are important in defining what is internationally accepted in terms of social justice. Many issues, including workers' rights, child labour and freedom of association demand clear international guidelines as organic food chains transcend local realities. Informing those who did not participate in their creation is a more instrumental process, although giving them meaning in a local context is not. Likewise, other issues (see Introduction and Social responsibility in the organic context) may be more amenable to local variations and interpretations. For these issues, a more emancipatory learning approach appears more appropriate. Collaboration and social learning at multiple levels throughout the supply chain between organic and social standard-setting bodies is key to addressing and improving this neglected piece of the organic movement.

Different kinds of learning about and towards social responsibility and at multiple levels are essential ingredients to improving social responsibility within the organic sector. Formation of coalitions, political alliances and innovative networks, as well as creative use of confrontation and conflict, would all make for many forms and levels of learning and lead to progress in the social responsibility arena. Collaborations like these within the sector need to be crystallized in agreed-upon and somewhat flexible regulations and economic incentives that support the sector and promote further learning. To reiterate our conclusions and recommendations, social responsibility in organic agriculture needs to include:

1. Standards and/or regulations on baseline social issues (see e.g., ILO, SAI, fair trade, IBS C, created and supported by multiple stakeholders in the organic food chain.
2. Collaboration between organic standard-setters and certifiers with other social standard-setters and certifiers (e.g., FLO, SAI, SAN, ILO, trade unions) to better address social responsibility.
3. Learning on social responsibility needs to be embedded at all levels within the organic supply chain (producers, processors, exporters, wholesalers, retailers) and certification hierarchy (standard-setting organizations, certification bodies, farmer organizations, farmers). Ideally the standards and regulations listed under a hierarchy are – temporary and always subject to revision – outcomes of such learning as opposed to the outcome of an instrumental process dominated by just a few powerful interests.

Fortunately, there has been much recent progress towards these goals via the innovative SASA Project and other ISEAL Alliance platform initiatives, IFOAM's Charter on Organic Trade and Code of Conduct for Organic Trade, and fair trade/organic certification collaboration in the field. The levels and kinds of learning presented are meant to provide opportunities and (organic) food for thought on how to stimulate the sector with respect to social responsibility and nourish the endogenous, local spirit of the organic movement. More research needs to be done to understand the interface and interplay between the three proposed mechanisms, if only to get a better grip on how to coordinate the three mechanisms and to create the proposed synergy between the three. The coordinating mechanisms and responsibilities within the entire chain need to become clearer in order to foster social responsibility throughout the organic sector that is grounded in social learning and is supported by a dynamic framework of standards and regulations.

11

Research and Development in Agriculture

Research Advances Organic Farming and Food Production

Agriculture and Organic Farming in particular are developing rapidly, not only as a result of technological change but also to changes in agricultural policy and public expectation. Research allows new knowledge to be developed and is thus vital for the future of Organic Farming. The question, What is the purpose of research on organic farming? Is it to increase yield and productivity, to compare it with other forms of agriculture, or to quantify its environmental and social impact, minimize the adverse effects and maximize the benefits and so on? The answer is of course, that research has a role in all of these and many other aspects of developing the organic food chain. The precise purpose of the research is usually defined by the funding body, and may differ with whether the funding body is from the public or private sector. In this Chapter, development of organic research is examined briefly through both the public and private sectors and then some of the issues that surround research on Organic Farming in terms of approaches and appropriate methodologies is explored. The extent to which organic and conventional agriculture require different research and different approaches to research is explored.

Cropping aspects of organic production have received more attention than livestock aspects until recently; this is reflected in the number of refereed publications relate the later development of research on animal husbandry to the situation in Europe where the regulations for organic animal husbandry were established only in 1999, eight years after the regulations on crop husbandry. Biodynamic agriculture has received little funding from public sources, although considerable work has been carried out in some private research institutes. A short description of the development of biodynamic research and institutions in Europe and the USA. Reports of biodynamic systems in mainstream journals are relatively rare, although there are some notable exceptions.

Prospects for Organic Agriculture Farming

Every technology or interference in the natural ecosystem has its own side effect; from last half a century we are using the pesticides and herbicides indiscriminately to raise the crop production, but now its ill effect visible for us. Story of Malwa region of Punjab (cancer belt) and Kerala's endosulfan tragedy are epitome example of it. Now we achieve the self-sufficiency in food grain production our next motto to serve healthy food to people. We are looking for alternative nature based chemical free farming and that is organic farming. It also helps to get the premium price by exporting the chemical free agriculture commodity to the developed countries, day by day demand for healthy and residue free food increase all over the world as well as country itself. People are now ready to pay extra amount if they are ensured that they are purchasing genuinely raised food through natural system. Thus, strengthening the certification process in country is must.

It is definitely true that India had witnessed a tremendous growth in agricultural production in the era of green revolution. Food grain production, which stood at a mere 50 million tons at the time of independence, had increased almost five and half times to 273.38 million tons by the end of 2016-17 from 159.59 million hectares of cultivated area in country. The technologies involved during the inception of green revolution supported by policies and further propelled by agrochemicals, machinery and irrigation were the main driving forces for the enhanced agricultural production and productivity. Despite the fact that the food security of India was definitely addressed by these technologies, an important setback was that the farmers using these technologies were still had to depend upon the purchased inputs. With manufacturing of fertilizers and pesticides as the two major inputs of Green Revolution (GR) technologies, an important point of consideration was the need for fossil fuels and/or expensive energy which are associated with serious environmental and health problems. In last 50 years we are using heavy number of fertilizers and pesticides and we already reach on plateau and diminishing low of return start to work, so we need to apply more input (fertilizer and pesticides) to get small raise in production which cause second generation problem and few of such epitome examples are some regions of Punjab (cancer belt of country) and endosulfan story of cashew plantations area in Kerala. Insecticides and herbicides in ideal condition lethal for target group only, for non-target group and human it is safe but this principle is not followed strictly and indiscriminate use of these chemicals put human life and ecosystem health on verge. All these thing and unsustainability issue associated with modern agriculture force us to look back in history to know either we are not doing any mistake by depending on off farm inputs because crop production is a recycle system of nature by putting too much off farm input, we are making it fragile day by day. One of such natural, recyclable and sustainable approach of farming is Organic farming. It is the effective and cost-efficient way to achieve sustainable development in the agriculture sector (IFOAM, 2010). Organic source of nutrient also helps to combat with the problem of multi nutrient deficiency and low organic content in our soil which is affecting productivity of major food crops at farmer field.

Organic farming is a method of farming system which primarily aimed at cultivating the land and raising crops in such a way, as to keep the soil alive and in good health by use of organic wastes (crop, animal and farm wastes, aquatic wastes) and other biological materials along with beneficial microbes (bio-fertilizers) to release nutrients to crops for increased sustainable production in an eco-friendly pollution free environment. As per the definition of the United States Department of Agriculture (USDA), study team on organic farming – organic farming is a system which avoids or largely excludes the use of synthetic inputs (such as fertilizers, pesticides, hormones, feed additives etc.) and to the maximum extent feasible rely upon crop rotations, crop residues, animal manures, off-farm organic waste, mineral grade rock additives and biological system of nutrient mobilization and plant protection.

Standards for Organic Agriculture

The main principles of organic farming by are as follows:

1. To work as much as possible within a closed system, and draw upon local resources.
2. To maintain the long-term fertility of soils.
3. To avoid all forms of pollution that may result from agricultural techniques.
4. To produce foodstuffs of high nutritional quality and in sufficient quantity.
5. To reduce the use of fossil energy in agricultural practice to a minimum.
6. To give livestock conditions of life that confirm to their physiological need.
7. To make it possible for agricultural producers to earn a living through their work and develop their potentialities as human being.

The Main Pillars of Organic Farming

The main pillars of organic farming are as below:

1. Organic threshold standards.
2. Reliable mechanisms regarding certification and regulatory affairs.
3. Technology packages.
4. Efficient and feasible market network.

Environmental Impacts of Organic Farming

A high proportion of the Earth's land surface has been transformed by direct human action, with significant consequences for biodiversity, nutrient cycling, soil structure and biology, and climate. Agriculturally used land accounts for 38% of the world's total land area, of which 69% is permanent pasture, making it the leading type of agricultural land use; 28% is arable land and 3% is under permanent crops (FAO 2004).

The location and physical production conditions are important determinants of the types and intensities of agricultural land use; in particular, the management of relatively flat, fertile land has often been progressively intensified, with mechanization leading to increased field sizes, removal of boundary vegetation and increased application of agrochemicals. In contrast, traditional farming systems on marginal land, where possibilities for mechanization are limited because of steep or inaccessible terrain, are in many areas under considerable economic pressure or have been abandoned already. The term Organic Farming used in this chapter is based on the Codex definition of the FAO. Which relies on natural methods of building soil fertility and controlling pests and diseases, can be divided into two categories: certified organic farming and de facto organic farming. Whereas certified organic production is often oriented towards the market for food labeled as organic, the latter may represent the major part of organically managed land. Such de facto organic farming can often be found in resource-poor and/or agriculturally marginal regions where local populations have a limited engagement with the cash economy.

This includes many traditional farming systems found in developing countries, which have evolved through centuries to create agricultural systems adapted to local environmental and cultural conditions. This chapter covers the environmental impacts of organic farming in this wider sense and also includes agroforestry, a management system that integrates trees in the agricultural landscape and that is systematically applied in Organic Farming. Agroforestry is practiced in all agroclimatic zones but is most prevalent in the tropical belt. Organic Farming is generally perceived as a form of agriculture that is more favorable for the environment than conventional agriculture. In this chapter it is asked where, precisely, the differences are in environmental terms and whether the available empirical evidence allows for a generalization. Geographical differences and interrelationships are emphasized wherever relevant. Primary sources for this chapter are a review of research results, mainly for developed countries in temperate and subtropical zones. There is a lack of scientific research in developing countries in the tropics and subtropics, but from practical experiences and case studies, it may be assumed that results are transferable. Overview of the main fields of environmental impact examined and the specific aspects covered.

Area	Aspect/indicator
Biodiversity	Genetic diversity Floral diversity Faunal diversity Habitat diversity
Landscape	Landscape structures and aesthetic value
Soil	Soil organic matter (SOM) and acidity (pH value) Biological activity Soil structure Erosion Desertification
Ground and surface water	Nutrient use and balance Nitrate leaching Phosphorus Pesticides Pathogens
Climate and air	Carbon dioxide (CO_2) Nitrous oxide (N_2O) Methane (CH_4) Ammonia (NH_3)
Energy	Intensity of energy use Efficiency of energy use

A wide range of indicators are being used in the description of the environmental impacts of agriculture. Particularly, the Organization for Economic Cooperation and Development (OECD) pushes the development of a common set of indicators at an international level. The assessment presented here is based on the Driver-State-Response (DSR) framework that has been developed by the OECD in. This framework is internationally accepted and allows cross-country comparisons. Whenever appropriate, simplifications and modifications have been made.

Biodiversity

Since its beginnings, agriculture has been a source of positive and negative effects on the eco-system in terms of wildlife conservation and landscape. The level of intensity that modern agriculture has reached in wide areas of Europe has resulted in a decrease in biological diversity of domestic and wildlife species. The destruction of biotopes, as well as the simplifications of crop rotations, and the increasing input of synthetic fertilizers and pesticides have been responsible for changes in the habitat of many species.

Maintaining and enhancing biodiversity is considered central to developing a sustainable organic system. Organic farming depends upon stabilizing agroecosystems, maintaining ecological balances, developing biological processes to their optimum and linking agricultural activities with the conservation of biodiversity. Increased biodiversity improves and buffers ecological services such as pollination, pest control, maintenance of soil fertility, thus strengthening farming systems and practices. Building on that, some organic certification organizations have incorporated biodiversity requirements into their standards. The Swiss organic standards, for example, require farmers to use 7% of their land as semi-natural habitats.

While comparing the impacts of organic and conventional farming systems on biodiversity, identified the following problems:

- Variation in the definition of organic farming standards between countries;
- Disparity between studies in their control for extraneous variation;
- Variation in the period the studies were carried out;
- Variation in the spatial scale; and
- Use of different measures of biodiversity.

Genetic Diversity

Today, the adoption of high yielding, uniform breeds and varieties has led to a considerable reduction in the number of species and in the number of varieties/breeds within species used in agriculture. There are many schemes and projects worldwide working to conserve seed banks and indigenous varieties, many of which are linked to Organic Farming projects. A typical example is the Sustainable Agriculture and Rural Development Project (SARDI) in Kenya where a community indigenous seed conservation program is being implemented. Indigenous seeds have been shown to perform better in drought conditions. Indians in the High Jungle of Peru who organically cultivated cotton and thereby conserved ecotypes adapted to the moist tropical conditions. Vreeland emphasizes that native cotton can also provide valuable genetic materials for improving commercial cotton varieties because of the high degree of natural resistance to insects, disease and drought stress.

A possible future threat to genetic diversity and biodiversity in general, could be the side effects of the release of genetically engineered or modified organisms (GMOs) into the environment (Soil Association 2001). In Organic Farming genetic engineering is banned.

Floral Diversity

In general, the diversity of floral species is closely connected to local site conditions. Nevertheless, while investigating the flora of arable and mixed farming, recorded higher weed abundance and species richness in fields under organic management in almost all studies. In several studies, organically managed fields held considerably rarer and/or declining species. However, in regions with low potential for biodiversity, the positive impact of organic farming on wild herb or grassland diversity is less distinct.

Factors that contribute to greater floral diversity include the proportion of grassland relative to arable cropping, the variation in sowing dates for cereal crops and the inclusion of both autumn- sown and spring-sown cereals. A richer floral diversity has positive impacts on faunal diversity, because it offers overwintering sites, refuges and areas with network links to other habitats. Weeds are considered competitive to the crop in conventional farming and are eliminated by herbicides, whereas in organic systems, some of the accompanying plants are desired to a certain degree and are considered useful as they provide ecological services. In the context of pollinators, which greatly benefit from a diversity of flowers, flowering weeds are more diverse and abundant in organic arable fields and grassland compared to conventional fields.

In organic grassland, the average number of species was higher than in conventional, as a result of lower stocking rates and lower fertilization levels in organic farms. Additionally, the mowing date is often delayed, which means that grass species can reach the flowering stage and thus achieve greater reproduction, leading to plant communities rich in species and structure.

Faunal Diversity

The effects of organic farming on faunal biodiversity have been studied, particularly for soil fauna and birds. In most cases, organic farming displays a greater faunal biodiversity than conventional farming. Key factors are: greater fauna-friendly crop protection management, the organic fertilization regimes, the more diversified crop rotation, and the more structured landscapes with semi-natural habitats and field margins. Landscape structures are essential for the survival of many invertebrates, especially due to favorable food and overwintering conditions. They also function as habitat cross links between meadows, fallows and field margins.

Earthworms

Earthworm populations can indicate the structural, microclimatic, nutritive and toxic situations in soils. They are highly suitable bioindicators of soil fertility, and they are known for their sensitivity to synthetic pesticides and to many agricultural practices. Earthworms help to improve soil structure and provide a high concentration of nutrients in a form accessible to plants. Many investigations in Europe and North America show that organically managed soils tend to exhibit a greater abundance and species number of worms compared to conventionally managed plots or farms.

Arthropods

Pesticides can affect beneficial arthropods either directly via contamination, or through alteration of the microhabitat and a reduction of their prey. In contrast to conventional farming, synthetic pesticides are banned in organic farming and the use of a few approved natural pesticides is very restricted. Most natural pesticides are not very selective and can therefore also affect beneficial arthropods. But the toxicity of natural pesticides is usually lower and their degradation significantly faster in comparison to synthetic pesticides. Additionally, the higher fertilization levels in conventional fields lead to a greater density of crops, which can alter the microclimate and can also reduce the occurrence of species dependent on a warm microclimate and on light demonstrated that tomato plants fertilized with high nitrogen (N) levels showed a higher intrinsic rate of increase of greenhouse whitefly populations. Studied the effects of organic and synthetic fertilizers on pests (aphids) and predatory arthropods associated with tomatoes. The application of both organic and synthetic fertilizers could increase pest populations. However, they found lower populations of aphids on organically fertilized tomatoes after one year, indicating that organic fertilizers may have the potential to reduce pest attacks in the long term.

Birds

The decline in farmland bird populations is well documented. Birds are well-suited indicator organisms that show the environmental status of nature and landscape infrastructure including agricultural land. Several studies show that bird densities are higher in organic farms. A study by Rhône-Poulenc (1997) has shown a steady annual increase in the number of bird territories on land converted to organic production and a larger overall number of territories on organically managed land. The British Trust for Ornithology (1995) found that breeding densities of skylarks and densities of birds in general were significantly higher on organic farms. In coffee production in Latin America, another factor appears more important research carried out by the Smithsonian Migratory Bird Centre in Colombia and Mexico showed more than 90% fewer bird species in sun-grown coffee plantations as opposed to shade-grown coffee. Shade-grown practices are recommended in organic standards as they fulfill requirements to enhance soil fertility, pest and disease control and expand crop production options.

Mammals

There are only two strictly comparative studies of organic and conventional farming and their influence on mammalian biodiversity. Activity levels of small mammals (wood mouse, bank vole, common shrew) were greater in organic than conventional fields. Increased food abundance through hedgerows and field edge habitats was cited as the most significant factor that benefitted these and a range of other mammalian species occurring on farmland. Many small mammal species are likely to have been affected by reduced insect and weed-seed food resources. Investigated bat activity and species richness in a paired (organic/conventional) farm. Total bat activity was significantly greater on organic farms than on conventional farms, while no significant difference in species richness was found between the farm types. Responsible factors for the greater activity levels appeared to be better water quality and greater prey availability.

12

Sustainable Farming Economic Management

Financial Management

Many in Organic Farming have moved on from whether Organic Farming works to how it works, and finding efficient solutions for if and when it does not work. Already in the 1980s, there was a movement away from comparing farm results between the two systems towards exploring different options on an organic farm. Despite this resolution, comparisons continued in many countries, if only to provide a reference to changes under organic management.

No attempt is made here to write an exhaustive treatise of the economics of Organic Farming. Rather, the aim is to highlight key issues affecting returns to management in Organic Farming, often comparing them with returns to management on a conventional farm. Production, input costs, output prices, premiums and subsidies are discussed to arrive at the farm profitability on established organic farms.

Productivity and Output

Productivity is production related to inputs. Productivity is often equated with production (e.g., tonnes of grain, head of cattle) per unit of land (hectare), but productivity could also be measured as output per unit of labour, energy or water used, or investment. Many assume that agricultural production per hectare under organic management is lower than under conventional management. Although this is often the case, it is not necessarily so. There are some general rules that can be observed.

Yield changes (changes of production per hectare or animal) are often relative to the intensiveness of the enterprise before conversion. This is true within enterprises (e.g., changes on intensively farmed cereals are larger than on extensively farmed cereals after converting to organic management) and between enterprises. For example, changes in yields in intensively run horticultural enterprises may be different from those in intensive cereal–livestock farming after conversion. In countries with a relatively high percentage of agriculture in extensive industries, such as dairy farming as compared with crops only, more farmers tend to convert to Organic Farming (e.g., in Austria and Switzerland), possibly reflecting the ease of conversion within this industry compared with other industries.

As the intensity of agriculture in a certain industry, say cereal–livestock, differs between countries, figures for changes in yield due to a change in management system also differ between countries. For example, reductions in crop yields after conversion were up to 40% in some European countries such as the UK, Germany, Denmark and the Netherlands in the 1990s, while those in Australia, Canada and the USA were between 10% and 20% down and were, in some cases, even higher than those under conventional farming mentioned yield decreases of 20% to be normal in organic systems, grain yield decreases of between 30% and 40% in several European countries.

For dairy, yield changes are more similar between countries. A reduction of milk yields of 30% to 40% per hectare. Mention a decrease of up to 20% in European countries per cow per year, and between 20 and 40 of cows per hectare. In Australia, recorded a decrease in milk liters per hectare of between 30% and 35% on seven pairs of farms. In Canada, found average milk sales per cow on 7 organic farms to be 4% down compared with 111 conventional dairy farms, and those per hectare were 41% lower. In the USA, average milk sold per cow was 13% lower on six organic farms compared with 27 conventional enterprises in the same region, while no figures on stocking density were provided. In developing countries, where the choice for Organic Farming is often made in the context of no-use of pesticides and fertilizers, the United Nations Food and Agriculture Organization (FAO) mentioned that organic management may even increase yields. Very few data are available for developing countries.

The expense of Production

Yield is not the only indicator of farm productivity. Inputs used to deliver those yields should also be considered, and can be in the form of materials, as discussed below, but they may also be in different forms. For example, one way to manage soil fertility and pest problems on organic farms is to employ a different rotation from that on conventional farms where synthetic fertilizers and pesticides can perform those tasks. Because of the difference in rotation, which may also mean a larger diversity of crops, the whole farm needs to be considered when determining productivity and profitability. For the farmer, the net profitability of the farm is important (i.e. what is left of the total receipts when all costs have been paid). For the profitability of the farm, the input costs can be as important an issue as the quantity produced and the prices received for the production. In many agricultural enterprises, organic farmers may use fewer inputs such as fertilizers and pesticides and sometimes more labour, such as for hand-weeding, than on conventional farms. This is not necessarily the case, as materials to manage nutrients and pests on organic farms such as mineral fertilizers, compost and pheromones can also be costly. Total variable costs to be typically 50% to 60% lower for organic cereals and grain legumes; 10% to 20% for potatoes and horticultural crops; and 20% to 25% for dairy cows, mainly due to reduced concentrates. For pigs and poultry, the extra cost of organic feed often means no reduction in average input prices. Labour (classified as fixed costs) is seen to be mainly higher on organic farms, while other costs including power, repairs, depreciation of machinery, property charges and depreciation and capital costs were reported as similar in most countries, though higher in Germany and lower in Australia. However, in a later survey among Australian cereal–livestock farmers, this difference was much less pronounced than in the earlier study, or non-existent. Labour cost is often controversial as it differs greatly between industries. For example, it is more likely to be similar or higher in intensive industries (e.g. horticulture, where labour can contribute considerably to replace herbicides) than in extensive industries (e.g. large-scale cereal growing). However, there can also be variations within industries.

Production Costs and Premiums

In relation to price premiums, it is important to discuss the more recent studies, as this variable is rather dependent on time. On premium prices on wheat, milk and beef between 1994 and 1997 in several European countries. For wheat, typical premiums were between 50% and 200%; for dairy, between 8% and 36%; and for beef, premiums tended to be between 20% and 30%. For the last two categories, the average premiums were low, at least partly because some of the product was sold on the conventional market without premiums. Estimated that premiums for cereals in the EU countries in 2001 reached just over 100%. For oilseeds, the premiums were recorded as 70%, for potatoes 166% and for wine 32%. Average price premiums of all organic goods to be 80%. Several products scored over 100%, such as whole meal flour, muesli, olive oil, spaghetti (the highest was 287%), several vegetables (beans, zucchini, and carrots), hard cheese and minced beef. However, not all organic products are sold on the organic market, and percentages of produce receiving a premium can vary considerably. For example, the range in Australia was estimated to be between 10% (sheep meat) and 95% (fruit and vegetables) being sold in the organic market in 2000 to 2001, with almost three-quarters of the total grains sold as organic, two-thirds of beef, and half of the organic milk supply.

Profitability

A general picture emerges of lower yields and productivity per farm, lower input costs and higher output prices on organic farms than those on conventional comparisons. The results in the different studies included in ranged from lower net returns in the studies in the USA, UK and Switzerland, similar in Australia and higher in studies from Germany, Denmark and Wales. Inclusion of premiums is, of course, an important determinant of the results. In some countries, such as those in Europe, premiums were more common than in others. In the Australian study, for example, very few farmers received premiums at the time of the first survey in the mid-1980s. In a later study, all farmers received premiums though the net farm returns were variable, being similar to those of conventional neighbors in one year and considerably lower in the next. Obviously, factors other than price premiums may be of greater significance in the total picture. In several different studies on both arable and dairy farms in ten European countries, net profits per hectare on organic farms were mainly within 20% of conventional farms. When the measure was profit per family work unit, the results for organic farmers were found to be similar or better than for conventional farmers. Profits varied considerably both by locality (country) and type of enterprise. Arable organic farms in particular were doing well, though specialized; highly intensive farms were generally not profitable under organic management. Which were important contributory factors in the profit figures. Not many data were available for horticultural, pig and poultry farms, possibly indicating that few farmers had converted to organic management in those industries. In a time series analysis of data for five countries (Switzerland, Germany, Austria, the Netherlands, Denmark) between 1990 and 2001, this same study showed mainly higher profits per hectare on organic farms. Annual results on organic and conventional farms follow very similar patterns, indicating that outside factors such as climate, prices and policies are rather more important for the profitability of the farm than the difference between the two systems.

In the dairy study in Australia, biodynamic farms showed a considerably lower net return over the three years surveyed than their conventionally farming neighbors, though their returns were similar compared with all farmers of the area. These results were achieved with no or low premium prices for the biodynamic milk. To reach similar returns to those of their conventional neighbors, the biodynamic farmers would have needed a premium price of about A$0.10 per liter of milk, a premium not considered extravagant in some other countries. The study in Canada showed a 12% lower net income per hectare on organic farms, though the return was 1 percentage point higher for the organic farms as returns on total assets. The USA study showed 16% lower net income per liter of milk produced.

Conclusion for Balanced Farms

Looking at the different studies over time on the comparability of organic and conventional management systems, the results of the net returns and their components of yield and total production, input costs and output prices are variable. In some studies net returns were found to be lower, and in others similar or higher. The question of comparative profitability is unlikely to be answered definitively, as the results depend on many factors, only some of which are within the farmer's control. The management of the farm and decisions about which crop to grow where, or even which enterprise to include, are somewhat under the farmers control. Climate and soil type, input and output prices, premiums and subsidies depend on the particular country and time in which the farmer lives, with policies changing over time. Perhaps, the inclination of the organic movement to minimize comparisons of the system in the 1980s was reasonable, especially if they were carried out to convince the conventional sector that investment in this system was warranted, rather than to find where the change in management system impacted most, and how to make the system work better.

In the debate about whether it is financially rewarding to farm organically, the question of what to include in the final analysis remains. The private benefits to the farmer are included. From society's point of view, another relevant factor is the difference in off-farm effects between the two management systems. This is notoriously difficult to quantify but is still relevant for the full picture of the total efficiency of the systems. In some of the studies discussed, this factor is already included in the form of subsidies and premium prices, though these extra revenues must also cover the extra cost of small-scale marketing. So, the answer is not clear cut. The studies discussed here indicate that the results can be positive for organic farmers, but do not necessarily need to be so. When considering conversion, the farmer would be wise to take several factors into consideration, including the physical attributes of the farm and the farmer, marketing possibilities and the political climate in the country at that time.

13

Challenges with Conversion

Whether a farmer can become an established organic farmer and not go bankrupt in the process depends on several factors. Important in this connection is the capacity of the farmer to adapt to a different way of management, and the ability to consider availability of farm resources on the history of the fields, market possibilities and political climate. This last issue involves government support in the form of, for example, production subsidies and research funding or reducing hurdles through regulations that hinder organic production and marketing. Devoted several chapters to the effect of conversion in several countries, and quote several early studies. Whatever the physical and financial problems of transition to Organic Farming may be, possibly one of the most difficult to overcome is the misconceptions about whether Organic Farming could work, and if it does, how. Lack of information about what changes are needed, and how to implement those changes, has always been mentioned as a major problem. Managerial ability, being able to manage the farm in a way appropriate to the system (e.g. handling different varieties, changing seeding dates to beat pests, recognizing problems before they cause great damage, and having knowledge about solutions) is an important first requirement. In the first years of moving towards organic management, most farmers have some special physical problems that they are unlikely to experience once the system is established. A biological change towards a different equilibrium in soil biota is one conversion aspect that may take time. Soil microorganisms essential for organic management need to become established. For example, Australian organic grain farmers reported that as they did not burn straw under the organic system, they experienced problems during transition with planting and germination of seed due to the straw blocking up the machinery, necessitating adaptation of their planting machinery. However, after those first few years the straw did break down considerably faster, presumably as a result of build-up of fungi over time. The absence of specific biological activity may also be part of the cause of yield reductions reported by some farmers in early years compared with yield levels further into the conversion period.

A change in farm layout is often advisable when changing management systems, and would require resources. For example, changes in rotations and livestock use within the rotation may require extra capital expenditure for fencing and livestock purchase. However, on farms where livestock require concentrates that are sourced off-farm, the change may mean some form of decrease in stocking rate.

Extra farm storage to cope with a change in marketing strategy can be another reason why more investments are needed. Buyers of farm produce may not have separate storage space for organic products, although this problem may decrease over time.

At the same time as the issues of change need attention and action, farmers can usually not count on premiums for the products. In the first phase of organic management, a farmer can be certified only as being in conversion. This means that during this time, premiums are less certain than at a later stage when the farm is certified as fully organic. The length of this period differs between countries, but is typically between one and three years. Whatever problems arise, the farmer needs to plan well ahead, and analyze possible scenarios under conversion that encompass changes, including the use of existing resources, the need for investments, possible changes in yields and total productivity, availability of labor and machinery, accessibility to markets and likely output prices and cash flow.

Risk and Uncertainties

Changing from conventional to Organic Farming means taking risks and facing different uncertainties. The risk during conversion is made up of the factors just discussed (see Conversion challenges), changes to the soil biota resulting from farming system changes, a steep learning curve for farmers regarding general organic farm management with particular emphasis on the characteristics of each particular farm, and a possible need for more investments in machinery, fencing, livestock and storage while outputs and marketing may be in flux. These risks can be expected to decrease over time as the physical processes settle down and the farmer learns more about the management system requirements. However, differences in production variability may be a more permanent feature of the two management systems. Yields on organic farms that are closer to the average in extreme years are mentioned in the early literature, but not always found. One study that did find a correlation between organic management and decreased variability of wheat yields was carried out in Australia, where extremes in weather conditions are not uncommon. Almost a decade later, concluded that in plots with two organic systems at the Rodale Institute in the USA, yields were higher in four of the five drought years, while the fifth year gave mixed results. The authors attribute the difference in yields under extreme conditions between the systems to the higher water-holding capacity of the soils in the organic plots. Variability in output prices for organic products depends on local arrangements, with possibilities ranging from fixed prices for a particular period, to a fixed percentage or amount on top of conventional prices, to a totally deregulated market. Variability in income resulting from changes in farm output prices is determined more by outside factors than by organic management.

Huge Economic Conversions

What happens when many farmers move towards Organic Farming? This question usually centers on the assumption of collapsing premiums, leading to lower farm incomes. These studies must make assumptions about changes in, for example, future input and output prices, yields, subsidy and premium levels, the way consumers and producers will react to changes in prices (price elasticity of demand and supply), population increase or decrease and level of adoption. The possible outcomes are too numerous to summaries here, so only a few results are presented as examples. Farm incomes are the variable considered most often in this respect, but they are not the only variable affected by a large-scale change to organic farming. As organic management requires a change in rotation for soil fertility and pest management reasons, a change of management system by many farmers would make a substantial change to the land use of a whole country. Hence, there will be a change in the relative importance of total output of different products and in output prices at the national level, and effects on farm incomes at state and individual levels. Other factors not discussed here that are sometimes analyzed include food security, employment and income in related industries, environmental and social benefits and public expenditure.

The direction and magnitude of the changes in production and returns will differ between countries. In the estimated that with a 10% increase in Organic Farming, there would be a decrease in wheat, barley, potatoes, sugar beet, oilseed and livestock and an increase in oats and field beans. In Australia, on large-scale cereal livestock farms, large scale conversion would lead to decreased total production of all cereals (wheat, oats, barley, canola), and an increase in sheep. The same Australian study estimated total farm incomes under the assumption of an adoption rate of 30%. Under the worst-case scenario, where premiums for crops were assumed to decrease from 15% to 0% with no premiums for livestock products at all stages (extreme assumptions) total returns to the sector would drop by 7% when 30% of farmers had converted. In the best-case scenario, with premiums decreasing from 15% to 7.5% (probably more realistic), total returns to the cereal livestock sector would have a 3% decrease at the 30% adoption rate. Many studies have been based on the assumption that a very large proportion of the farming sector becomes organic often 100%. However, this is rather unlikely, and would certainly not happen under conditions that are easily imaginable. More likely is that when, say, 10% or 20% of farmers have converted, political pressure may result in, for example, more research relevant to organic farming, information systems more readily accessible to organic farmers, decreased consumer prices as a result of decreased marketing costs (economies of scale) with a resulting increase in demand for organic food, and confidence of conventional farmers to convert. The detailed study on the Danish agricultural sector assumed initially that 80% of farmers converted to Organic Farming, and also analyzed what happened at lower levels of conversion. The results showed that with prevalent input and output prices, and assumptions of reducing premiums with an increasing number of farmers across all industries, the total receipts to the farming community would start to fall only after about 25% of farmers had moved to organic farming. If that situation were reached, the political and social climate within the agricultural sector would have changed substantially, warranting new estimations with assumptions closer to reality than those available now.

In summary, studies have been undertaken to examine the consequences of large-scale changes towards organic management among farmers. Total production is usually the main factor considered, although total returns to farming and to farmers were also analyzed. Studies mentioned here indicate that conversion to organic management, especially when the conversion rate is not taken to extremes, would generate few changes that cause adjustment problems to the agricultural sector.

The market for organic produce has developed considerably. In 2000, the global retail market was estimated to be around US$16 billion, with an estimated annual growth in demand of about 15% to 20% in several markets. Estimates for 2001 were closer to US$20 billion, and forecasts for 2003 for the world organic food and beverages market were between US$23 billion and US$25 billion (ITC 2005). Organic food and non-food sales grew by about 20% during 2003, and reached US$10.8 billion in one of the main markets, the USA (Organic Trade Association 2004). The ITC estimate for the organic retail market reached US$30–32 billion in 2005. A general upward trend can be detected, which shows that both the demand and supply of organic products are growing.

Possibilities and Advantages, Substantial collection of Information and expertise

Organic farming has become more popular since the 1980s. This growth has been stimulated by, and has stimulated, an increase in know-how and managerial skills on farms. Although globally there may not yet be many people who have a profound knowledge of Organic Farming, experience in many kinds of applicable disciplines such as farm management (farmers), soil science, plant pathology, entomology, veterinary science, plant and animal husbandry, marketing, economics and political and social science is now available.

Standards and accreditation in Numerous Nations

For farmers to be able to sell their products, consumers require sufficient confidence in the organic production and marketing processes to be willing to buy. Over the last 20 years, considerable effort has been expended by both public (state authorities) and private (e.g., farmers associations) organizations into developing a well-functioning certification scheme. As a result, many countries have credible organic standards and certification procedures.

Market Accessibility

The demand for organic products has increased especially during crises in conventional agriculture. Examples include mad cow disease since the late 1980s, the presence of dioxin in feed in Belgium in 1999, and foot and mouth disease in the UK in the early 2000s. The increase in demand in many places, especially Japan and China, should provide future opportunities for organic production.

Threats and Weaknesses maintaining Standards and Certification

Just as Organic Farming benefits from food scandals in the conventional market, it also suffers when a scandal occurs in the organic market, for example with producers, certification offices and through accidental problems. With organic produce commanding premium prices, unscrupulous conventional farmers may be tempted to sell their produce on the organic market. This is a potential problem especially in markets where the word organic is not legally defined. In Australia, for example, the potential for two problems arises because standards cover requirements for the export market, but the term organic has not been defined for the domestic market. The first is the possibility of fraud in the domestic market, where there is little possibility of recourse; and other problems could occur in the import market where, due to World Trade Organization rules relating to national treatment, the Australian government cannot prohibit imports and sales of non-certified products labeled as organic. Second, certification organizations may be tempted to cut corners. In a market where all produce needs to be certified and certification organizations vie for business, such behavior is conceivable in a bid to provide the cheapest certification service. Third, accidents can also be problematic. In mid-2002, poultry feed on more than 100 German organic chicken farms was contaminated with the pesticide nitrogen. The feed had been stored in a warehouse previously used for pesticide storage.

International trade standards and certifications for Organic Farming have developed over time. Originally, many countries, or organizations within countries, developed standards and systems suitable for their local circumstances. These were not necessarily based on scientific principles, or principles that were accepted in the scientific community. Although such problems are slowly being addressed; many remain. The differences between standards and compliance systems among countries can make international trade problematic for exporters because exporting countries need to comply with the requirements of the importing countries. As the main importing countries have different standards and certification requirements, complying with the requirements of the importer can become costly. This issue is one that the International Task Force set up by the United Nation's Conference for Trade and Development (UNCTAD), the FAO and the International Federation of Organic Farming Movements (IFOAM) has been trying to solve since 2002. The problems can be daunting for farmers in many countries, especially developing countries that do not have their own certification schemes.

The problem of different organic standards and certification for imports is compounded in some countries by private certification schemes that convince local supermarkets to accept only products certified by the particular private certification scheme. A similar effect is achieved by local certifiers promoting produce from local producers, such as the UK Soil Association, which mentions in the Organic Action Plan a target of 70% of organic primary products to be sourced from the UK. Thus, not surprisingly, the contention of some exporting countries is that organic import requirements aim to keep competition out and benefit local producers rather than to provide consumers with a guarantee that the product was organically grown, which was the main aim. The issue of regional standard appropriateness is not controversial. The problem is, however, to find efficient ways to facilitate international trade in the presence of regional standards and certification procedures.

The issue of acceptable differences in standards is also an area that needs more attention, as many standards are not based on scientific considerations. One example is the conversion period, which determines the time farmers have to wait before they are eligible to be called organic.

This provides easier access to premium prices. At the international level, the conversion period is set in different ways. For example, IFOAM has a conversion period of 12 months for plant production, while the EU and the Codex Alimentations Commission (part of the FAO and World Health Organization's food and veterinary standards activities) require a minimum of between two and three years, although the conversion period can be reduced to one year if certain requirements are met. Importing countries that set the conversion period at a certain level do not take into account the possibly legitimate claim by some exporters that a decrease in conversion for their local circumstances may be appropriate. For example, if the conversion period is meant to provide time for the breaking down of substances that have no place in Organic Farming or for positive agents to build up, this process will be influenced by local conditions, such as climate. Variable conversion periods in countries with different climates would be appropriate.

Purchasing Costs

Apart from the importance of consumer confidence in the product, which is encouraged through a trustworthy and known certification scheme, a stable supply (e.g., through the involvement of supermarkets) and relatively low prices are important for Organic Farming to expand. In general, produce sold as organic commands higher prices than conventionally grown produce of equal quality. Many surveys have shown that consumers are willing to pay more for these products, though the demand is highly price sensitive. Demand rapidly shifts towards conventional products with increasing prices for organic products. Though output premiums for farmers are often blamed for high consumer prices, for many products the farm price is only a small part of the final consumer price. Other components such as transport, wastage, processing, handling and sales also influence the retail price. One would expect that for many products with low sales volumes, costs would be relatively high per unit; however, the lack of institutional facilities for organic products is also likely to affect costs. This has been recognized by for example the Dutch and Danish governments who devote a relatively large part of their Organic Farming support in encouraging and promoting the increased efficiency of the supply chain for organic products.

Federal Laws and Regulations

Differences in government support between countries can lead to an unfair advantage for producers in countries with regulation compared with those where no regulations exist. For example, subsidies for organic farming in one country, be it in the form of direct payments for farmers, research subsidies, or the development of a national certification scheme, can affect the price and the quantity of production in that country. This means that producers in other countries will need to produce more efficiently in order to be able to export to those countries. In other words, an advantage for farmers in some countries means a disadvantage for some farmers in other countries. Some governments, especially those of EU countries, support Organic Farming.

Subsidies have an additional benefit to increasing farmers' income farmers are able to sell their produce more cheaply, thereby reducing consumer prices and therefore demand. However, the potential for price reduction is likely to be tempered by increased input costs, especially those of land. In this chapter, the economics of Organic Farming for farmers is considered, which includes issues surrounding yields, inputs, outputs and returns to farming. Off-farm effects are not considered here. In general, the financial results can be positive for organic farmers, but are not necessarily so. Details about the local situation are important for the results, such as the history of the farm, the particular enterprise, prevailing input and output prices, and domestic and international policies. The question also arises how easy it is to get to the state of organic management without failing financially in the first (transitional) stage. Important factors for success include planning for conversion related to the use of existing resources, the need of investments, changes in yields and total production, input availability and prices, marketing opportunities and cash flow. As whole-country transition to organic farming would influence such measures as input availability and prices and output prices, studies have been undertaken to model the effects of such changes. No disasters are obvious, though studies heavily depend on assumptions.

For the future, one of the strengths of the industry can be found in an increasing body of knowledge, not only on organic practices, but also on the effect of policies. Other strengths include the established standards and certification systems, and the existing market. Weaknesses or threats seem more numerous. The first is related to standards, including fraud, both by producers and certification agencies, and the possibility of accidents. The second is in the area of international trade, where the proliferation of standards and compliance schemes can make trade difficult for exporting countries. The desire of some countries to restrict international trade and the absence of a scientific approach in some of the criteria are two more issues in the arena of international trade. Other threats concern consumer prices and governmental regulations. High consumer prices are a deterrent to growth in demand of organic products, and government regulations in one country cause a non-level playing field for producers in other countries.

14

Overall Principles of Biodynamic Agriculture

Biodynamic agriculture can be described as a comprehensive agricultural system on mixed farms, which should always involve crops and livestock. The system is based on respect for, and efforts in, gaining awareness of the spiritual dimension of all creatures and the inorganic environment. Insights into this spiritual dimension (e.g., detailed descriptions of the spiritual character of animals, plants and physical elements or of planetary impacts) were given by Rudolf Steiner in 1924 (Steiner 2004). These descriptions should give farmers an orientation in their actions and awareness in order to evolve the agricultural system. Furthermore, the formulation of specific preparations was given by Steiner as a particular new element in agriculture. Biodynamic agriculture incorporates the landscape and the ecosystem as essential parts of the whole. It minimizes nutrient inputs from outside the farm through proper legume crop management.

Fundamentals of Biodynamic Farming

Globally, about 3270 holdings in 35 countries and with an area of 104,000 ha are certified under the Demeter label. This means that they comply with both the relevant national organic farming standards and the international Demeter Standards (Demeter International 2003). There are a huge variety of enterprises on Demeter farms, covering anything from temperate arable farming to viticulture in France, cotton production in Egypt and silkworm breeding in China. The intentionally individual design of life processes by the farmer, as determined by site conditions, is one of the basic tenets of biodynamic agriculture. This principle emphasizes that humans have a responsibility for the development of their ecological and social environment which goes beyond economic aims and the principles of descriptive ecology. The vision on which this principle is based is the individual design of the holding in the context of the complex interaction of all impacting factors. The pedosphere, ecosphere and landscape, as well as the atmosphere and the cosmic environment (apart from the sun, these are primarily the moon and the planets) form the natural basis. Crop plants, livestock, the farmer as well as the entire socioeconomic environment have an effect at all levels of this natural environment, are influenced by these levels, and thus form an intricate interrelationship. A farm, therefore, becomes an 'individuality' in which the various factors, just like organs, have specific functions and are interlinked through feedback relationships. Biodynamic farming takes account of these interrelationships and tries to actively shape them into guided processes through a variety of management practices.

The prime objective is always to encourage healthy conditions for life. Soil fertility, plant and animal health, as well as product quality are to be maintained and continuously improved in a largely closed system by designing the individual farm enterprises and operations accordingly. The following principles are based on the above objective. They represent implicit targets of the biodynamic method which are to be implemented on the holdings (Demeter International 2003). The Demeter International trademark indicates that produce is grown and processed in compliance with their standards. Not all biodynamic farms or gardens are associated with Demeter as an organization. For farms producing under the Demeter trademark, the following principles are compulsory. However, for other farms producing biodynamically, these principles are guides, but are not always binding.

Diverse holdings

Arable farming together with livestock is a prerequisite of biodynamic management. There are special regulations for horticultural enterprises and perennial crops. A variety of enterprises that balance out each other's one-sidedness are seen as a prerequisite for a healthy farm organism. The keeping of livestock, especially ruminants, is an essential element. Management largely aims at closed nutrient cycles. The use of mineral nitrogen (N) fertilizers is prohibited. Application of mineral calcium, potassium, phosphorous and magnesium fertilizers conforming to the Demeter Standards is permitted. It has been repeatedly shown that such management practices can indeed be sustainable.

Varied site-appropriate Crop Rotations

A well-planned crop rotation is the basis for healthy plant growth and a fertile soil. Where crop rotations cannot be practiced (perennial crops e.g., fruit or grapevine), strips of cover crops with flowering plants are sown. A rhythmic alternation of nitrogen-fixing (e.g., legumes) and nitrogen-demanding plants (e.g., Brassicas), shallow-rooted (cereals) and deep-rooted plants (lucerne), soil-building (root crops) and soil-exhausting crops (cereals), or humus-building (fodder plants) and humus-destroying crops (root crops) utilizes the effect of each preceding crop, enhances the formation of the soil structure and prevents crop rotation diseases. Legume cultivation for feed, as a grain legume or as green manure, is an indispensable part of the crop rotation (25-30%) in order to replace the required nitrogen. An ideal crop rotation maintains a year-round green cover, improving soil fertility in the most sustainable manner, and preventing nutrient losses.

Intensive Organic Fertilization

Fertilization does not aim primarily at feeding the plants but at enlivening the soil. This basic tenet of biodynamic farming requires intensive fertilization using farmyard manure, composts, liquid manure, green manure and others. The use of farm-generated fertilizers varies with the crop grown and is targeted at growing high-quality foods. For example, potatoes are given farmyard manure; vegetables receive only well-rotted composts. Organic fertilizers are treated with biodynamic compost preparations (see Biodynamic preparations) as standard practice in order to enliven the fertilizer as well as the soil. Regular applications of organic fertilizers serve to close and intensify nutrient cycles.

Biodynamic Preparations

To support and intensify the processes in the organic manure and in the soil, Steiner indicated six different preparations that should be added together to the manure. Furthermore, introduced two field spray preparations that are used to support fertility and plant development at certain growth stages (Steiner 2004, 4th lecture). The compost preparations (formerly preparations 502-506) are all made from medicinal plants: yarrow flowers (Achillea millefolium), chamomile flowers (Chamomilla recutita), whole plant of stinging nettle (Urtica dioica), oak bark (Quercus robur), dandelion flowers (Taraxacum officinale) and valerian flowers (Valeriana officinalis). In most cases, these are stored in certain animal organ 'sheaths' (intestine, peritoneum, bladder) and fermented for a certain time buried in the soil or under the influence of air. The two field sprays are cow manure ('horn manure', formerly preparation 500) and quartz meal ('horn silica', formerly preparation 501), both buried in the soil in cow horns, horn manure during the winter, horn silica during the summer. After they have been dissolved in water and stirred rhythmically for one hour, these substances are sprayed on plants and soil.

Livestock husbandry systems which fulfill the animals' welfare requirements

Livestock husbandry, which is based on regarding the animals as beings with soul's results in stricter standards than are normally applied. For example, the dehorning of cows is prohibited; breeding for isolated performance criteria is avoided and the aim is to breed for longevity and lifetime yield. In this type of livestock husbandry ruminants, for example, are regarded as animals whose role is to utilize roughage. Breeding aimed at high milk yields, which can only be sustained with high concentrate inputs, is considered adverse to the animals' welfare requirements. From the point of view of maturity, heifers for example, should be given sufficient time to develop prior to being put in calf for the first time.

Food quality – the stuff of life

The production of high-quality food for people and animals is an implicit aim of the biodynamic method. The question as to how food quality can be defined is challenging for both practitioners and researchers, and has led to intensive development on quality assessment methods. Great importance is attached to differentiation and ripening processes in the production and processing of foods.

Breeding locally adapted Crop Plant Varieties

Only biodynamic farming currently has its own organic plant-breeding programs, which aim to provide varieties that are suited to biodynamic cultivation and are efficient in their management. This field of action is a key issue for biodynamic agriculture. In terms of breeding objectives, the priorities are plant health, suitability of the respective environmental conditions, the ability to save seed and propagate them on-farm, as well as high food quality and yield.

Maintaining Plant Health

Starting with the use of plant protestants and management agents in accordance with biodynamic standards, biodynamic farming uses further specific preparations. Spraying horsetail tea (Equisetum arvense), in addition to the 'stimulating' preparations, to protect crops from fungal diseases. For pest regulation, the ashes of burnt specimens of the pest species are to be used.

Cosmic Rhythms

The annual rhythms and the rhythms of the moon, as well as those of the planets, are considered in sowing, planting, and harvesting and sometimes also for fertilizer applications.

Low-impact Soil Management

Soil cultivation measures are predominantly carried out with the aim of enhancing soil fertility and soil processes, and controlling for weeds. Depending on location and crop type, biodynamic farming uses all types of soil cultivation from ploughing to no-till agriculture.

Including the Landscape

Landscape design as an element of farm management has always been considered in biodynamic farming. There is a growing awareness of the role of landscape design at the farm level.

Design of the Social and Economic Setting

The concept of land ownership is increasingly being questioned on biodynamic farms and the land is no longer necessarily regarded as private property. Often it is transferred to charitable trusts or similar organizations. Ways that aim to build economic relationships between producers and consumers based on the principles of mutual responsibility and respect are being pioneered. Agricultural work is also regarded as a therapeutic opportunity, and mentally handicapped or spiritually ill people, troublesome youngsters, addicts, and convicts are often integrated into the farm communities. The training of apprentices is of high priority, and in many countries, there are comprehensive training courses or agricultural colleges that specifically teach biodynamic farming methods. Regular courses and seminars for working farmers are another field that is given much attention.

These guidelines leave much scope for each individual farm to work out its own strategy congruent with the living conditions in its surroundings. The farm individuality, to which much importance is attached, results from this detailed design. The manner of soil cultivation, the exact design of crop rotations, fertilization regimes, choice of varieties, the apportionment of livestock categories, the balance between enterprises, the nature of social relationships on the farm and economic relationships with the customers are all-incumbent on the farmers and are not regulated. However, they are documented as part of the national Demeter inspections and are discussed and evaluated in professional journals, at regular association meetings and at an annual international conference in Switzerland.

Background – what are the roots of biodynamic farming?

The concepts underlying biodynamic agriculture relate in particular to the spiritual radiance of all earthly phenomena, including that of the inorganic pedosphere and the cosmos. The physical–sensory reality is seen as an expression of the diversely structured spiritual world, the development of which takes place through the processes in the physical world. However, this idea of development is not strictly teleological, as it regards human beings as active creators of future developments both in the positive and the negative sense. Therefore, the biodynamic concept of ecology is not a conserving but a developing one. In the world has its cause in a spiritual world that is actually present and can fundamentally be an object of human awareness. Anthroposophy, insofar as it concerns the perception of nature, understands itself as giving an additional dimension, the spiritual one, which in no case would replace or deny any basic natural law. The way to develop the ability to attain knowledge of this spiritual world through the awareness of the sensory world, and the manner in which this spiritual world acts and appears, was described by Steiner in many books and lectures. The biodynamic farming system was founded in June 1924, when Steiner met with about 60 farmers in Koberwitz near Wroclaw (Poland) and gave eight lectures about a modern agriculture based on spiritual science. Today these eight lectures still form the essential and formative basis of biodynamic farming, with much practical and scientific experience during the past 80 years having more closely defined and evolved many aspects of this foundation. Steiner gave these lectures at the request of farmers who were involved in the development of anthroposophy, and when he gave the Koberwitz lectures, in agreement with his listeners, he based his deliberations on previous lectures.

Research - what are the main questions faced by biodynamic farmers and researchers?

Biodynamic research is often associated with research on the preparations and cosmic influences. In the early days, much scientific attention was indeed given to the preparations. Many attempts have been made to provide evidence for the effect of the preparations on soil parameters, plant development, root growth intensity and food quality (IBDF 2003). Results vary widely. However, most of the studies assess the overall impact of the biodynamic method, of which the use of the preparations is but one part.

Some studies, including more prominent ones, have repeatedly shown a clearly positive impact of the biodynamic method on soil structure, enzyme activities, CO_2 exchange and earthworm populations. However, there are also studies that were not able to show such results. With regard to plant development, the aim is to achieve a compensating, stabilizing effect that levels out extremes, and these effects have repeatedly been observed. The impact of cosmic rhythms on plant growth has been verified many times; however, no conclusive explanation of the reasons for the variation in the results has yet been produced.

Essential research focuses on plant breeding, appropriate breeding objectives for livestock, the veterinary treatment of livestock, ruminant feeding with low concentrate inputs, food quality and food quality evaluation methods. It is necessary to drive forward research on the effectiveness of the biodynamic preparations in order to provide sound practical recommendations for the frequency and timing of their use. However, discussions are required to decide whether it makes sense to assess the effect of individual biodynamic practices in isolation from the overall method, as these practices were proposed as part of a comprehensive farming system, the impulse of which is based, in particular, on a holistic understanding of agricultural processes. A complex, holistic, systemic form of science would be appropriate to biodynamic farming, but has not yet been realized. It is unclear what would constitute holistic research. Phenomenological descriptions of the overall context are without doubt a key element that has been developed in much detail in the context of anthroposophic research over the past 80 years.

Index

P

R

S

V

Z

◆ ◆ ◆ ◆ ◆